An Algebraic Geometric Approach
to Separation of Variables

Konrad Schöbel

An Algebraic Geometric Approach to Separation of Variables

 Springer Spektrum

Konrad Schöbel
Friedrich-Schiller-Universität Jena
Germany

Habilitationsschrift Friedrich-Schiller-Universität Jena, 2014

ISBN 978-3-658-11407-7   ISBN 978-3-658-11408-4 (eBook)
DOI 10.1007/978-3-658-11408-4

Library of Congress Control Number: 2015949632

Springer Spektrum
© Springer Fachmedien Wiesbaden 2015

Springer Spektrum is a brand of Springer Fachmedien Wiesbaden
Springer Fachmedien Wiesbaden is part of Springer Science+Business Media
(www.springer.com)

An Algebraic Geometric Approach to Separation of Variables

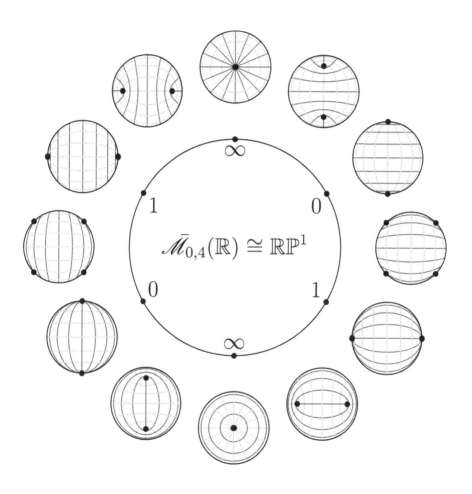

To Alba & Aitana

Preface

Separation of variables is one of the oldest and most powerful methods to construct exact solutions of fundamental partial differential equations in classical and quantum physics, like the Hamilton-Jacobi equation in Newtonian mechanics, the Schrödinger equation in quantum mechanics or the wave equation. A separation of the Schrödinger equation for the hydrogen atom in spherical coordinates, for example, yields functions describing the orbital structure of the electrons, i.e. the basis for the periodic table of the elements, and is thus at the root of chemistry. Likewise, many other well known special functions which are used all over in science and technology stem from a separation of variables.

The problem to classify all coordinate systems in which this method is applicable, the so-called *separation coordinates*, was solved exhaustively by Ernest G. Kalnins & Willard Miller Jr. over 30 years ago. For this reason many experts would consider the theory of separation of variables as settled or even old-fashioned. However, in this book we argue that the above classification problem is essentially an algebraic geometric and not a differential geometric problem, i.e. governed by algebraic instead of partial differential equations. This means that Kalnins & Miller's list of separation coordinates carries a much deeper geometric structure, namely that of a projective variety equipped with the natural action of the isometry group G. From a categorical point of view the classification problem has been solved in the category of sets, but not in its natural category, the category of projective G-varieties.

It seems that, albeit obvious, this fact has so far been completely overlooked (or ignored). Virtually nothing is known about the geometry of these varieties or the topology of their quotients. The aim of the present book is to bridge this gap and to lay the foundations for a consequent algebraic geometric treatment of variable separation. By applying it to spheres, we not only give a proof of concept that the approach we propose is viable, we also demonstrate that it leads to surprising results already for this simplest family of constant curvature manifolds. Namely, we reveal a correspondence between two a priori completely unrelated objects: the space of equivalence classes of separation coordinates on the n-dimensional sphere $\mathbb{S}^n$ and the Deligne-Mumford moduli space $\overline{\mathcal{M}}_{0,n+2}(\mathbb{R})$ of stable algebraic curves of genus zero with $n + 2$ marked points. Moreover, we derive a classification of separation coordinates via Stasheff polytopes from this correspondence, together with a simple and uniform construction based on the natural operad structure on the family of moduli spaces $\overline{\mathcal{M}}_{0,n}(\mathbb{R})$.

In this way we build a bridge between the theory of separation of variables, whose origins date back far into the 19th century, and most recent results in modern algebraic geometry.

Mathematics subject classification: *Primary 58D27; Secondary 53A60, 14D21.*

Keywords: *separation of variables, Killing tensors, algebraic curvature tensors, Stäckel systems, Deligne-Mumford moduli spaces, stable curves with marked points, associahedra, Stasheff polytopes, operads*

Contents

3 The generalisation: a solution for spheres of arbitrary dimension   99

4 The perspectives: applications and generalisations   121

Acknowledgements   131

Bibliography   133

0 Introduction

Contents

0.1 Separation of variables

Separation of variables is one of the most powerful methods for solving partial differential equations and one of the very few general methods that yield exact solutions. The idea is to seek for a solution which is a product (or a sum) of functions, each of which only depends on a single variable. The strength of this ansatz lies in reducing a partial differential equation in n variables to n differential equations in only one variable, since the theory of ordinary (single-variable) differential equations is far better developed than the theory of partial (multi-variable) differential equations.

It turns out that separation of variables is only possible in certain coordinate systems. It is thus a classical problem to classify all coordinate systems in which one can solve a given partial differential equation by this method – the so called *separation coordinates.*

Most well known exact solutions for fundamental equations in nature stem from a separation of variables: the Hamilton-Jacobi equation in classical and semi-classical mechanics, the Schrödinger equation in quantum mechanics, the Poisson equation in electrostatics and Newtonian gravitation, the heat equation in thermodynamics or the wave equation. Separating variables for these equations in the different types of separation coordinates is the origin of the defining ordinary differential operators for a plethora of special functions, which are used all over in mathematics and theoretical physics, far beyond the theory of separation of variables.

0.2 History

19th century. Separation of variables is also one of the oldest methods for solving partial differential equations, with its origins dating back to the year 1837, when Gabriel Lamé introduced this method in order to separate the heat equation and describe thermodynamic equilibrium configurations for homogeneous ellipsoids [Lam37]. The first geometric characterisation of separation coordinates was given in 1891 by Paul Stäckel in his *Habilitationsschrift*, where he showed that every orthogonal system of separation coordinates on an n-dimensional Riemannian manifold gives rise to what now is called a *Stäckel system* [Stä91]. In modern language, this is a system of n linearly independent commuting Killing tensors that satisfy a certain integrability condition.

Stäckel's result turns integrable Killing tensors into the fundamental object for the study of separation of variables.

20th century. In 1934, Luther P. Eisenhart completed Stäckel's work by showing the converse, i.e. that every Stäckel system arises from orthogonal separation coordinates [Eis34]. His result established a one-to-one correspondence between orthogonal separation coordinates and Stäckel systems and thereby paved the way for a classification of orthogonal separation coordinates by differential geometric means. In the same publication Eisenhart used this correspondence to classify orthogonal separation coordinates in dimension three for the Euclidean space and the sphere. Later Ernest G. Kalnins and Willard Miller extended his classification to all constant sectional curvature manifolds and to arbitrary dimension, by giving a rather sophisticated graphical procedure from which the separation coordinates can be deduced [KM86, Kal86].

Since the pioneering works of Stäckel and Eisenhart, an extensive amount of literature has been published on separation of variables: not only on various generalisations, such as non-orthogonal separation, complex separation or so called R-separation, but also on related concepts like separable potentials, multiseparability, superintegrability, special function theory or symmetries of differential operators. All these works have in common that their methods remain within the realm of Riemannian geometry, mostly working with local coordinate expressions. This observation is rather surprising, given that the algebraic geometric nature of the problem should have been apparent at least since the 1950ies. Indeed, the space of Killing tensors K is a linear space and the Nijenhuis form of the integrability conditions is a non-linear system of partial differential equations for K which, albeit very complicated, is algebraic in K and its covariant derivative ∇K [Nij51].

21st century. Perhaps the only exception are recent works that adopt a view on the classification of separation coordinates from the perspective of Geometric Invariant Theory (GIT), by trying to find isometry invariants of Killing tensors that are capable of classifying orthogonal separation coordinates [HMS05]. This approach had not become viable until recently, because it relies on an extensive use of modern computer algebra. In dimension three for example, the algorithm requires the solution of a linear system of approximately 250,000 equations in 50,000 unknowns and the complexity of the algorithm renders it impractical in dimensions greater than three [DHMS04]. We are not aware of any applications of the resulting invariants to elucidate the algebraic geometric structure of the corresponding GIT quotient.

0.3 Aim

The present book deals with the classification of orthogonal separation coordinates on constant sectional curvature manifolds, with emphasis on spheres. The reader may well wonder why this should still be considered a topic of interest, since the theory is more than 175 years old and Kalnins & Miller gave an exhaustive solution over 30 years ago. But the meaning of the word "solution" here depends on the category one is working in. Notice that Kalnins & Miller give a *list* of separation coordinates. That is, from a categorical point of view, they describe the space of orthogonal separation coordinates in the *category of sets*. But, as follows from the Nijenhuis form of the integrability conditions, it is obvious that this space carries a natural structure of a projective variety, equipped with a canonical action of the isometry group. The same is true for the space of integrable Killing tensors.

It seems that nobody ever considered separation of variables from this algebraic geometric point of view. To the best of our knowledge, virtually nothing is known about the algebraic geometry of these varieties or the topology of their quotients under isometries. That is to say that in the *category of topological spaces* as well as in the *category of projective varieties*, the classification problem for separation coordinates has never been worked out. The overall aim of the present book is to bridge this gap and to give a detailed description of the algebraic geometry of the space of integrable Killing tensors and the space of orthogonal separation coordinates as well as their quotients under isometries. The precise mathematical formulation of this aim will be made clear in the next section.

We can only speculate why this has never been undertaken before. The reason is probably that the algebraic equations involved seemed far too complicated for a direct solution. Interestingly, Stäckel already commented on this in his *Habilitationsschrift* from 1891, where he notes [Stä91, p. 6]:

> Die Diskussion dieser Gleichungen ergab, dass es für $n = 2$ drei wesentlich verschiedene Formen der Gleichung $H = 0$ giebt, bei denen diese notwendigen Bedingungen erfüllt sind, und da diese Gleichungen auch wirklich Separation der Variabeln gestatten, ist die Frage für den Fall $n = 2$ vollständig erledigt. Aber schon für $n = 3$ werden die algebraischen Rechnungen so umständlich, dass mir eine weitere Verfolgung dieses Weges aussichtslos erschien. [1]

In a sense, the aim of the present *Habilitationsschrift* is to accomplish Stäckel's computations for arbitrary n, by means of representation theory, algebraic geometry and a substantial progress made in the theory of moduli spaces in the last few decades.

---

[1] "For $n = 2$ the discussion of these equations resulted in three essentially different forms of the equation $H = 0$ [the Hamilton-Jacobi equation, note from the author], for which the necessary conditions are satisfied, and since these equations indeed allow a separation of variables, the question is completely settled in the case $n = 2$. However, already for $n = 3$ the algebraic computations become so cumbersome, that it seemed hopeless to me to pursue this approach further."

0.4 Mathematical formulation

0.4.1 Separation coordinates

We say that the Hamilton-Jacobi equation

$$\tfrac{1}{2} g^{ij} \frac{\partial W}{\partial x^i} \frac{\partial W}{\partial x^j} = E$$

separates in local coordinates $x^1, \dots, x^n$ on an n-dimensional Riemannian manifold if it admits a solution of the form

$$W(x^1, \dots, x^n; \underline{c}) = W_1(x^1; \underline{c}) + \dots + W_n(x^n; \underline{c}),$$

depending on n parameters $\underline{c} = (c_1, \dots, c_n)$ with

$$\det \left(\frac{\partial^2 W}{\partial x^i \partial c_j} \right) \neq 0.$$

Note that if we reparametrise each coordinate x^i with a strictly monotonic function Φ_i, the Hamilton-Jacobi equation is still separable in the new coordinates $\Phi_i(x^i)$. The same is true for a permutation of the variables. In order to avoid this arbitrariness, we consider different coordinate systems as equivalent if they are related by such transformations. By abuse of language we will call a corresponding equivalence class simply *separation coordinates*. Equivalently, we can think of separation coordinates as the (unordered) system of coordinate hypersurfaces defined by the equations $x^i =$ constant. The separation coordinates are called *orthogonal*, if the normals of these hypersurfaces are mutually orthogonal.

0.4.2 Killing tensors, Nijenhuis integrability and Stäckel systems

The main tool in studying orthogonal separation coordinates are Killing tensors that satisfy a certain integrability condition.

Definition 0.1. *A* Killing tensor *on a Riemannian manifold M is a symmetric form $K \in \Gamma(S^2T^*M)$ satisfying*

$$\nabla_u K(v, w) + \nabla_v K(w, u) + \nabla_w K(u, v) = 0, \qquad (0.1)$$

where ∇ is the Levi-Civita connection of the metric g on M. We denote by $\mathcal{K}(M)$ the linear space of Killing tensors on M.

Note that the metric g is trivially a Killing tensor, because it is covariantly constant.

Remark 0.2. *In the above definition, a Killing tensor is a symmetric bilinear form. But we can use the metric to identify this symmetric bilinear form with a symmetric endomorphism. Depending on the context, we will adopt one or the other interpretation.*

We say a symmetric endomorphism K is *integrable*, if around any point in some dense open set we can find local coordinates such that the coordinate vectors are eigenvectors of K. This geometric definition can be cast into an analytic form with the aid of the Nijenhuis torsion of K, defined by

$$N(X, Y) := K^2[X, Y] - K[KX, Y] - K[X, KY] + [KX, KY]$$

or, in local coordinates x^α on M, by

$$N^\alpha{}_{\beta\gamma} = K^\alpha{}_\delta \nabla_{[\gamma} K^\delta{}_{\beta]} + \nabla_\delta K^\alpha{}_{[\gamma} K^\delta{}_{\beta]}. \qquad (0.2)$$

Integrability in the above sense is then equivalent to the following system of nonlinear first order partial differential equations [Nij51]. We take this characterisation as the definition of integrability.

Definition 0.3. *A symmetric endomorphism field $K^\alpha{}_\beta$ on a Riemannian manifold is* integrable *if and only if it satisfies, in local coordinates x^α on M, the* Nijenhuis integrability conditions

$$0 = N^\delta{}_{[\beta\gamma} g_{\alpha]\delta} \qquad (0.3a)$$

$$0 = N^\delta{}_{[\beta\gamma} K_{\alpha]\delta} \qquad (0.3b)$$

$$0 = N^\delta{}_{[\beta\gamma} K_{\alpha]\varepsilon} K^\varepsilon{}_\delta, \qquad (0.3c)$$

where the square brackets denote complete antisymmetrisation in the enclosed indices.

Definition 0.4. *A Stäckel system on an n-dimensional Riemannian manifold is an n-dimensional space of Killing tensors which*

(i) are integrable.

(ii) mutually commute with respect to the commutator bracket

$$[K, L] = KL - LK. \tag{0.4}$$

Note that the metric is contained in every Stäckel system.

It can be shown that every Stäckel system contains a Killing tensor with simple eigenvalues on an open and dense subset [Ben93]. By definition, the n distributions given by the orthogonal complements of its eigendirections are integrable. Hence they define n hypersurface foliations with orthogonal normals or, equivalently, orthogonal coordinates. On the other hand, every Killing tensor commuting with the above and having simple eigenvalues defines the same coordinates. In this manner each Stäckel system defines a unique coordinate system. It is a classical result that these are separation coordinates and that every system of orthogonal separation coordinates arises in this way from a Stäckel system.

Theorem 0.5. *[Stä91, Eis34] Locally, there is a bijective correspondence between Stäckel systems and orthogonal separation coordinates.*[2]

The significance of Stäckel systems goes beyond separation of variables, since they provide an important class of examples for dynamical systems which are completely integrable in the Liouville sense. In this context the Killing tensors are quadratic first integrals of motion.

0.4.3 Observations

Let us start with three simple observations:

---

[2]see the comment on the locality of this result on page 15

- The Killing equation (0.1) is linear, so the space of Killing tensors is a linear space.
- The covariant derivative ∇K depends linearly on K.
- The Nijenhuis integrability conditions (0.3) are (homogeneous) algebraic equations in K and ∇K.

Combining them results in the following important observation:

Observation I. *The set $\mathcal{I}(M) \subseteq \mathcal{K}(M)$ of integrable Killing tensors on a Riemannian manifold M carries a natural structure of an algebraic variety, a subvariety in the vector space $\mathcal{K}(M)$ of Killing tensors on M.*

This variety possesses two remarkable structures coming from its geometric origin. One is the action of a symmetry group, the other a kind of fibration by Stäckel systems. The latter is not a fibration in the proper sense, since all Stäckel systems meet in the subspace spanned by the metric and two different Stäckel systems may share a bigger subspace. Our next observation is a consequence of the fact that the commutator bracket (0.4) is obviously algebraic in its arguments. Recall that a Stäckel system is, by definition, an n-dimensional subspace in the space of Killing tensors.

Observation II. *The set $\mathcal{S}(M) \subseteq \mathcal{G}_n\big(\mathcal{K}(M)\big)$ of Stäckel systems on an n-dimensional Riemannian manifold M carries a natural structure of a projective variety, a subvariety in the Grassmannian $\mathcal{G}_n\big(\mathcal{K}(M)\big)$ of n-dimensional subspaces in the space $\mathcal{K}(M)$ of Killing tensors on M.*

Regarding natural symmetries, let us make three more simple observations:

- The Killing equation is invariant under the commuting actions of the following two groups:
  - Isom(M): the isometry group, acting on $\mathcal{K}(M)$ by pullback
  - Aff$(\mathbb{R}) = \mathbb{R} \rtimes \mathbb{R}^*$: the affine group, acting on $\mathcal{K}(M)$ by
  $$K \mapsto bK + ag \qquad (a, b) \in \mathrm{Aff}(\mathbb{R})$$

- The Nijenhuis integrability conditions (0.3) are invariant under these groups.
- The vanishing of the commutator bracket (0.4) is invariant under these groups.

This brings us to our last observation.

Observation III. *The algebraic variety $\mathcal{I}(M)$ is naturally equipped with two commuting group actions, induced by the actions of the isometry group* $\mathrm{Isom}(M)$ *and the affine group* $\mathrm{Aff}(\mathbb{R})$ *on the space of Killing tensors. The projective variety $\mathcal{S}(M)$ is equipped with a natural action of the isometry group.*[3]

0.4.4 The problems

Observation I naturally leads to the following problem for a given Riemannian manifold M.

Problem I. *Determine the algebraic variety $\mathcal{I}(M)$ of integrable Killing tensors on M. More precisely:*

(i) Give explicit (homogeneous) algebraic equations defining $\mathcal{I}(M)$.
(ii) Solve these equations explicitly.
(iii) Give an algebraic geometric interpretation of the variety $\mathcal{I}(M)$.
(iv) Find polynomial isometry invariants characterising the integrability of a Killing tensor.

Remark 0.6. *The Nijenhuis integrability conditions (0.3) yield infinitely many algebraic equations for $K \in \mathcal{K}(M)$, one for each point $x \in M$ and choice of tangent vectors $u, v, w \in T_x M$. By Hilbert's basis theorem these can be reduced to a finite set. So Problem (i) asks to do this explicitly.*

In view of a classification of separation coordinates we are not so much interested in single solutions of the Nijenhuis integrability conditions, which are points on the variety $\mathcal{I}(M)$, but rather in

---

[3]The affine group acts trivially on $\mathcal{S}(M)$.

Stäckel systems, which are maximal subspaces of commuting solutions. Together with Observation II this leads to the following problem.

Problem II. *Give an algebraic geometric interpretation of the projective variety $\mathcal{S}(M)$.*

For a classification we also want to regard separation coordinates as equivalent which differ by any of the obvious symmetries. Together with Observation III, this leads to the following problem.

Problem III. *Let $G = \mathrm{Isom}(M) \times \mathrm{Aff}(\mathbb{R})$ be the product of the isometry group of a Riemannian manifold M with the affine group of $\mathbb{R}$.*

(i) Describe the quotient $\mathcal{I}(M)/G$ of the variety of integrable Killing tensors under the group G.

(ii) Identify the Stäckel systems in $\mathcal{I}(M)/G$.

(iii) Describe the quotient $\mathcal{S}(M)/\mathrm{Isom}(M)$ of the variety of Stäckel systems under isometries.

(iv) Find polynomial isometry invariants classifying Stäckel systems (and thereby separation coordinates) under isometries.

Note that $\mathcal{S}(M)/\mathrm{Isom}(M)$ uniquely parametrises Stäckel systems up to symmetries or, which is the same, equivalence classes of orthogonal separation coordinates. So this is the *classification space* we are ultimately interested in.

We have formulated the above problems in full generality, since they make sense for any (even pseudo-)Riemannian manifold M. In what follows we limit ourselves to constant sectional curvature manifolds and spheres in particular. We will solve Problems I and III for the 3-dimensional sphere as well as parts (ii) and (iii) of Problem III for spheres of arbitrary dimension. Although from the latter a formal description of the variety $\mathcal{S}(M)$ can be derived, a satisfactory solution to Problem II is still missing.

0.5 Method

The key result on which our approach is based is an explicit identification of the space of Killing tensors on constant curvature manifolds as a representation space for the isometry group. Recall that any non-flat (pseudo-)Riemannian manifold of constant curvature is locally isometric to a hypersurface

$$M := \{x \in V : g(x,x) = 1\} \subset V, \qquad (0.5a)$$

isometrically embedded into a vector space V equipped with a non-singular bilinear form g. For positive definite g this is the round sphere and if g has Lorentzian signature we obtain hyperbolic space. Other signatures yield pseudo-Riemannian manifolds such as de Sitter or anti-de Sitter spaces. The isometry group of M is simply the (pseudo-)orthogonal group $O(p,q)$, where (p,q) is the signature of g.

Most of the results we are going to prove for curved constant curvature manifolds will hold for a flat space as well, if it is embedded as the affine hyperplane

$$M := \{x \in V : g(x,u) = 1\} \subset V \qquad (0.5b)$$

with unit normal $u \in V$.

Remark 0.7. *By abuse of language, we will refer to the two models (0.5), when we speak of constant curvature manifolds, although a general constant curvature manifold is only locally isometric to one of these.*

Killing tensors on constant curvature manifolds are described by algebraic curvature tensors in the following sense.

Definition 0.8. *An* algebraic curvature tensor *on a vector space V is an element $R \in V^* \otimes V^* \otimes V^* \otimes V^*$ satisfying the usual (algebraic) symmetries of a Riemannian curvature tensor, namely*

$$R(x,w,y,z) = -R(w,x,y,z) = R(w,x,z,y) \qquad (0.6a)$$

$$R(y,z,w,x) = R(w,x,y,z) \qquad (0.6b)$$

$$R(w,x,y,z) + R(w,y,z,x) + R(w,z,x,y) = 0, \qquad (0.6c)$$

called antisymmetry, pair symmetry *and* Bianchi identity *respectively.*

The following result is the starting point of our approach. It allows us to translate the Nijenhuis conditions and therefore the entire problem of separation of variables to the realm of algebra and algebraic geometry.

Proposition 0.9. *[MMS04] The vector space of Killing tensors K on a constant curvature manifold $M \subset V$ of the form (0.5) is naturally isomorphic to the vector space of algebraic curvature tensors R on the ambient space V. The isomorphism is explicitly given by*

$$K_x(v, w) := R(x, v, x, w) \qquad x \in M, \quad v, w \in T_x M, \qquad (0.7)$$

where we consider a point $x \in M$ as well as the tangent vectors $v, w \in T_x M$ as vectors in V satisfying $g(x, x) = 1$ respectively $g(x, v) = g(x, w) = 0$ (and similarly for flat space). This isomorphism is equivariant under the natural actions of the isometry group on Killing tensors respectively on algebraic curvature tensors, i.e. defines an isomorphism of representations.

Due to the above isomorphism, we will speak of the algebraic curvature tensor "corresponding to", "associated to" or simply "of" a Killing tensor on a constant curvature manifold.

0.6 Scope

We have made a number of assumptions and restrictions which we shall justify now.

Hamilton-Jacobi equation. Surprisingly, for the classification of separation coordinates it is not so important which of the classical partial differential equations one considers. The prototype equation is the Hamilton-Jacobi equation, because (additive) separation of this equation is a necessary condition for (multiplicative) separation of other equations like the Schrödinger equation, the Laplace equation or the Helmholtz equation. If the manifold has constant curvature, this condition is also sufficient [Eis34]. This is why we only mentioned the Hamilton-Jacobi equation when introducing separation coordinates.

Potential. So far we have neglected the fact that in its general form the Hamilton-Jacobi equation

$$\tfrac{1}{2}g^{ij}\frac{\partial W}{\partial x^i}\frac{\partial W}{\partial x^j} + V(x) = E \tag{0.8}$$

contains a potential function $V(x)$, taking account for an external force field. The reason to do so was that a necessary condition for the separability of (0.8) is the separability of this equation with $V(x) \equiv 0$. A sufficient condition is given by the so-called *Bertrand-Darboux condition*

$$d(KdV) = 0, \tag{0.9}$$

where K is any generic Killing tensor in the corresponding Stäckel system, i.e. one with simple eigenvalues, regarded as an endomorphism on 1-forms. The same is true for the Schrödinger equation.

Constant curvature. While the general theory is valid on an arbitrary Riemannian manifold and (with some care) even for pseudo-Riemannian manifolds, in this book we will first restrict to constant curvature manifolds and then to spheres in particular. The reason for restricting to constant curvature manifolds is threefold. First, they comprise the manifolds which are most important for applications in physics, namely flat Euclidean space, spheres, hyperbolic spaces, Lorentzian spaces and (anti-)de-Sitter spaces. Second, the models (0.5) in combination with Proposition 0.9 provide a simple realisation of these manifolds, their isometries as well as their Killing tensors. Third, as we will show, the results are already fairly involved for this simplest class of Riemannian manifolds and only very little is known for non-constant curvature. The reason for restricting to spheres later on is that their isometry groups, being compact, are the simplest amongst constant curvature manifolds.

Orthogonality. For constant curvature manifolds, the restriction to separation coordinates which are *orthogonal* does not constitute a loss of generality, because of the following result.

Theorem 0.10. *[KM86, Kal86] All separation coordinates on a constant curvature manifold are equivalent to orthogonal separation coordinates.*

This equivalence is given by a linear change in the so-called *ignorable coordinates*. These are coordinates which the metric does not depend on. We will consider separation coordinates up to this equivalence throughout this book.

Locality. A priori, Theorem 0.5 is only a local result. Nevertheless, any local Killing tensor field on a constant curvature manifold of the form (0.5) can be extended to a global one. Hence the same is true for the corresponding separation coordinates. That is why we can use said result for a global classification of orthogonal separation coordinates on constant curvature manifolds. Note, however, that the separation coordinates are only well defined on an open and dense set and become singular where none of the Killing tensors in the corresponding Stäckel system has simple eigenvalues.

0.7 Some examples

0.7.1 Cofactor systems

To give an example of Stäckel systems, let us consider the generic ones on a constant curvature manifold. They can be constructed from special conformal Killing tensors.

Definition 0.11. *A special conformal Killing tensor on a Riemannian manifold M is a symmetric form $L \in \Gamma(S^2 T^* M)$ satisfying*

$$\nabla_u L(v, w) = \lambda(v)g(w, u) + \lambda(w)g(v, u)$$
$$\lambda(u) = \tfrac{1}{2}\nabla_u \operatorname{tr} L. \tag{0.10}$$

The space of special conformal Killing tensors parametrises geodesically equivalent metrics, i.e. metrics having the same set of unparametrised geodesics. Their importance in our context stems from the fact that

$$K := L - (\operatorname{tr} L)g$$

defines an integrable Killing tensor (see Lemma 2.27). Thus, in the generic case where L (and hence K) has pairwise different eigenvalues, it defines a system of separation coordinates. The corresponding Stäckel system is spanned by the coefficients of the polynomial

$$K(\lambda) = \operatorname{Adj}(L - \lambda g) = \sum_{i=0}^{n-1} K_i \lambda^i,$$

where $\operatorname{Adj} L$ denotes the adjugate matrix, i.e. the transpose of the cofactor matrix of L [BM03]. That is why these systems are also referred to as *cofactor systems*.

We can deduce the corresponding separation coordinates directly from the special conformal Killing tensor L, since its eigenvalues are constant on the corresponding coordinate hypersurfaces [Cra03]. This means that the eigenvalues of L can be taken as coordinate functions. On a constant curvature manifold the situation is further simplified. The reason is that under certain conditions, which are met in this case, every special conformal Killing tensor L on M is the restriction of a covariantly constant symmetric form $\hat{L}$ on the metric cone over M and vice versa (see for example [MM10]). For spheres, the metric cone over $\mathbb{S}^n \subset V$ is nothing but V itself, so the determination of generic separation coordinates on $\mathbb{S}^n$ reduces to a computation of the eigenvalues of the restriction L of a constant symmetric form $\hat{L}$ on V.

0.7.2 Elliptic and polyspherical coordinates

To give an example of separation coordinates, let us consider the two extremal cases on spheres. The generic case of orthogonal separation coordinates on the sphere can be obtained from a special conformal Killing tensor L with simple eigenvalues as described above. For

$\hat{L}$ with (constant) eigenvalues $\Lambda_0 < \Lambda_1 < \ldots < \Lambda_n$ the eigenvalues $\lambda_1(x), \ldots, \lambda_n(x)$ of L at a point $x = (x_0, \ldots, x_n) \in \mathbb{S}^n$ are the solutions of the equation

$$\sum_{k=0}^{n} \frac{x_k^2}{\Lambda_k - \lambda} = 0, \qquad \|x\|^2 = 1 \tag{0.11}$$

which can be ordered to satisfy

$$\Lambda_0 < \lambda_1(x) < \Lambda_1 < \lambda_2(x) < \cdots < \lambda_n(x) < \Lambda_n.$$

This is nothing else but the defining equation for the classical *elliptic coordinates* on the sphere $\mathbb{S}^n$ introduced in 1859 by Carl Neumann [Neu59]. Note that shifting or multiplying the parameters $\Lambda_0 < \Lambda_1 < \ldots < \Lambda_n$ by a constant results in a mere reparametrisation of the same coordinate system. Therefore elliptic coordinates form an $(n-1)$-parameter family of separation coordinates on $\mathbb{S}^n$.

The other extreme, having no continuous parameters at all, are *polyspherical coordinates* as considered by Vilenkin [Vil65, VK93]. Each of these coordinate systems is given in terms of Cartesian coordinates by starting with $x(\varnothing) := 1$ on $\mathbb{S}^0 \subset \mathbb{R}^1$ and then defining recursively $z = z(\varphi_1, \ldots, \varphi_{n-1})$ on $\mathbb{S}^{n-1} \subset \mathbb{R}^n$ from $x = x(\varphi_1, \ldots, \varphi_{n_1-1})$ on $\mathbb{S}^{n_1-1} \subset \mathbb{R}^{n_1}$ and $y = y(\varphi_{n_1}, \ldots, \varphi_{n_1+n_2-2})$ on $\mathbb{S}^{n_2-1} \subset \mathbb{R}^{n_2}$ by setting

$$z = (x \cos \varphi_{n-1}, y \sin \varphi_{n-1}) \tag{0.12}$$

for $n = n_1 + n_2$. Since this involves a choice of a splitting $n = n_1 + n_2$ in each step, polyspherical coordinates on $\mathbb{S}^{n-1}$ are parametrised by planar rooted binary trees with n leaves.[4] For example, the standard spherical coordinates correspond to a left comb tree, i.e. a binary tree where each right child is a leaf.

---

[4]When Vilenkin introduced polyspherical coordinates in [Vil65], he used trees which are not binary. The description with binary trees appeared in Vilenkin & Klimyk [VK93, Chap. 10.5] and both are completely equivalent.

0.7.3 A trivial and a simple example: Spheres of dimension one and two

To give an example for the solution of the Problems I, II and III, let us consider the spheres of dimension one and two. The case $\mathbb{S}^1$ is trivial. The space of Killing tensors is 1-dimensional and spanned by the metric. Hence $\mathcal{K}(\mathbb{S}^1) = \mathcal{I}(\mathbb{S}^1) \cong \mathbb{R}$ and there is only a single Stäckel system, so $\mathcal{S}(\mathbb{S}^1)$ is a point. Taking quotients is trivial, because the isometry group leaves the metric invariant.

The case $\mathbb{S}^2 \subset \mathbb{R}^3$ is less trivial. Algebraic curvature tensors in dimension three have a zero Weyl part and are thus determined by their Ricci tensor alone, a symmetric bilinear form. That is, $\mathcal{K}(\mathbb{S}^2) \cong S^2\mathbb{R}^3$. In two dimensions the Nijenhuis integrability conditions (0.3) are void, because they involve an antisymmetrisation over three indices. This means that every Killing tensor is trivially integrable:

$$\mathcal{I}(\mathbb{S}^2) = \mathcal{K}(\mathbb{S}^2) \cong S^2\mathbb{R}^3.$$

Under the orthogonal group $O(3)$ every symmetric form on $\mathbb{R}^3$ is equivalent to a diagonal one, which is unique up to permutations of the diagonal elements. Consequently $\mathcal{I}(\mathbb{S}^2)/O(3) \cong \mathbb{R}^3/S_3$ and hence

$$\mathcal{I}(\mathbb{S}^2)/G \cong \mathbb{P}^1/S_3.$$

By definition, every Stäckel system on $\mathbb{S}^2$ is 2-dimensional and contains the metric. Hence it is a subspace in $\mathcal{K}(\mathbb{S}^2)$ spanned by the metric and some linearly independent Killing tensor. Linearly independent means, that we can choose its Ricci tensor to be trace free. Consequently, the projective variety

$$\mathcal{S}(\mathbb{S}^2) \cong \mathbb{P}S_0^2\mathbb{R}^3$$

is the (projective) space of trace free symmetric forms on $\mathbb{R}^3$ and we obtain[5]

$$\mathcal{S}(\mathbb{S}^2)/O(3) \cong \mathbb{P}^1/S_3.$$

This is the space visualised on the title page of this book.

---

[5]The isomorphism $\mathcal{I}(\mathbb{S}^2)/G \cong \mathcal{S}(\mathbb{S}^2)/O(3)$ is exceptional and false in higher dimensions.

0.8 Overview over the main results

Chapter 1 is the foundation of our algebraic approach, where we translate the Nijenhuis integrability conditions on a constant curvature manifold into purely algebraic equations and, using representation theoretic methods, cast them into the following simple form:[6]

Main Theorem I. *A Killing tensor on a constant curvature manifold $M \subset V$ is integrable if and only if the associated algebraic curvature tensor R on V satisfies, in any orthogonal basis of V, the following two conditions:*

$$\boxed{\begin{smallmatrix}a_2\\b_2\\c_2\\d_2\end{smallmatrix}}\, g_{ij} R^i{}_{b_1 a_2 b_2} R^j{}_{d_1 c_2 d_2} = 0 \qquad (0.13\text{a})$$

$$\boxed{\begin{smallmatrix}a_2\\b_2\\c_2\\d_2\end{smallmatrix}}\,\boxed{a_1\,|\,b_1\,|\,c_1\,|\,d_1}\, g_{ij} g_{kl} R^i{}_{b_1 a_2 b_2} R^j{}_{a_1}{}^k{}_{c_1} R^l{}_{d_1 c_2 d_2} = 0 \qquad (0.13\text{b})$$

Here the operators on the left hand side are the Young symmetrisers for a complete antisymmetrisation in the indices a_2, b_2, c_2, d_2 respectively for a complete symmetrisation in the indices a_1, b_1, c_1, d_1. For flat spaces, the bilinear form g on V has to be replaced by the (degenerated) pullback of the metric on M via the orthogonal projection $V \to M$.

This solves part (i) of Problem I for all (pseudo-)Riemannian constant curvature manifolds. In preparation of the solution to part (ii) of Problem III, we also translate the commutator bracket (0.4) between two Killing tensors into a purely algebraic expression for the associated algebraic curvature tensors (cf. Proposition 1.16).

---

[6]During the review process of the corresponding publication [Sch12] (submitted on 10 August 2010, accepted 15 January 2012), these equations appeared in the same form (with minor changes in notation) in an article by Cochran, McLenaghan & Smirnov, without neither a derivation nor a proper citation [CMS11] (compare Equations (0.13) here with Equations (3.32)–(3.34) there). In order to dispel any doubts on the provenance of these equations, the author would like to point out that his preprint [Sch10] (submitted on 16 April 2010) has been prior to the preprint version of said article [CMS10] (submitted on 22 September 2010).

Chapter 2 is the proof of concept for the feasibility of our approach, where we solve the algebraic equations (0.13) explicitly for the simplest non-trivial case – the 3-dimensional sphere $\mathbb{S}^3$. It would be too much to hope for the quotient space $\mathcal{I}(\mathbb{S}^3)/O(4)$ to be an algebraic variety. We will instead prove a result which is only slightly weaker: the existence of a slice for the isometry group action on $\mathcal{I}(\mathbb{S}^3)$. By a *slice* we mean here a linear section which meets every orbit in a finite number of points. This slice will be the space of algebraic curvature tensors that are diagonal in the following sense.

Definition 0.12. *Due to the symmetries (0.6a) and (0.6b), we can interpret an algebraic curvature tensor R on V as a symmetric bilinear form on $\Lambda^2 V$. We say that R is* diagonal *in an orthonormal basis $\{e_i : 0 \leqslant i \leqslant n\}$ of V, if it is diagonal as a bilinear form on $\Lambda^2 V$ in the associated basis $\{e_i \wedge e_j : 0 \leqslant i < j \leqslant n\}$. In components, this simply means that*

$$R_{ijkl} = 0 \qquad unless \quad \{i,j\} = \{k,l\}.$$

We denote by $\mathcal{K}_0(\mathbb{S}^n)$ the vector space of Killing tensors on $\mathbb{S}^n$ that have a diagonal algebraic curvature tensor (with resprect to some fixed orthonormal basis in V).

We can now state the result that is central to Chapter 2.

Main Theorem II.

(i) *Under the action of the isometry group any Killing tensor in $\mathcal{I}(\mathbb{S}^3)$ is equivalent to one in $\mathcal{K}_0(\mathbb{S}^3)$.*

(ii) *The variety*

$$\big(\mathcal{I}(\mathbb{S}^3) \cap \mathcal{K}_0(\mathbb{S}^3)\big)/\mathrm{Aff}(\mathbb{R}) \tag{0.14}$$

is isomorphic to the linear determinantal variety of matrices of the form[7]

$$M = \begin{pmatrix} \Delta_1 & -t_3 & t_2 \\ t_3 & \Delta_2 & -t_1 \\ -t_2 & t_1 & \Delta_3 \end{pmatrix} \qquad with \quad \mathrm{tr}\, M = 0 \tag{0.15a}$$

---

[7]Here the Δ_α's are the pairwise differences of the eigenvalues of the Weyl tensor and the t_α's parametrise the eigenvalues of the trace free Ricci tensor.

satisfying

$$\det M = 0. \tag{0.15b}$$

This describes a projective variety in $\mathbb{P}^4$ *which carries a natural* S_4*-action, given by conjugating* (0.15a) *with the symmetries of the regular octahedron in* $\mathbb{R}^3$.

(iii) Two Killing tensors in $\mathcal{I}(\mathbb{S}^3) \cap \mathcal{K}_0(\mathbb{S}^3)$ *are equivalent under the isometry group if and only if their corresponding matrices* (0.15a) *are equivalent under this* S_4*-action.*

Definition 0.13. *We will call the variety* (0.14) *the* Killing-Stäckel variety, *or* KS variety *for short.*

This already solves part (ii) of Problem I. Exploiting the algebraic geometric structure of the Killing-Stäckel variety then allows us to deduce the following results for the sphere $\mathbb{S}^3$.

Corollary 2.42: An algebraic geometric description of the space $\mathcal{I}(\mathbb{S}^3)$ of integrable Killing tensors on $\mathbb{S}^3$. This solves part (iii) of Problem I.

Corollary 2.12: A set of polynomial isometry invariants characterising the integrability of an arbitrary Killing tensor on $\mathbb{S}^3$. This solves part (iv) of Problem I.

Section 2.4: A juxtaposition between algebraic geometric properties of the KS variety on one side (such as singularities, projective lines and projective planes on the variety) and geometric properties of the corresponding Killing tensors on the other side. This solves part (i) of Problem III.

Proposition 2.23: An identification of the Stäckel systems in the KS variety. This solves part (ii) of Problem III

Propositions 2.36 and 2.38: An algebraic geometric description of the variety of Stäckel systems with diagonal algebraic curvature tensor and a topological description of the classification space for separation coordinates on $\mathbb{S}^3$. This solves part (iii) of Problem III.

Proposition 2.34: A set of polynomial isometry invariants classifying separation coordinates on $\mathbb{S}^3$. This solves part (iv) of Problem III.

In particular, we will prove that the space of Stäckel systems on $\mathbb{S}^3$ with diagonal algebraic curvature tensors is isomorphic to the blow-up $\mathbb{P}^2 \# 4\mathbb{P}^2$ of $\mathbb{P}^2$ in four points and that

$$\mathcal{S}(\mathbb{S}^3)/O(4) \cong (\mathbb{P}^2 \# 4\mathbb{P}^2)/S_4.$$

All material presented in the first two chapters is self contained, based only on Stäckel and Eisenhart's characterisation of orthogonal separation coordinates. We would like to emphasise that our approach does not rely on any computer algebra or numerical results, neither for the solution of the Nijenhuis integrability conditions nor for the construction of isometry invariants.

The thorough analysis of the case $\mathbb{S}^3$, as presented in Chapter 2, has provided just enough information to reveal the pattern for a generalisation to spheres of arbitrary dimension. Indeed, together with the two trivial examples from Section 0.7 we get the following sequence of spaces $\mathcal{S}(\mathbb{S}^n)/O(n+1)$ that classify separation coordinates on spheres: for $n = 1$ this is just a point, for $n = 2$ this is the quotient $\mathbb{P}^1/S_3$ and for $n = 3$ we obtained $(\mathbb{P}^2 \# 4\mathbb{P}^2)/S_4$. It turns out that the spaces $\mathbb{P}^1$ and $\mathbb{P}^2 \# 4\mathbb{P}^2$ form the first two non-trivial members of a very prominent family of projective varieties, namely the real Deligne-Mumford moduli spaces $\bar{\mathcal{M}}_{0,n}(\mathbb{R})$ of stable algebraic curves of genus 0 with n marked points, and that these spaces carry a natural action of S_{n-1}. This observation has lead to the statement of the following theorem, which is the principal result of Chapter 3.

Main Theorem III. *[SV15] The Stäckel systems on $\mathbb{S}^n$ with diagonal algebraic curvature tensors form a nonsingular algebraic subvariety of the Grassmannian $G_n(\mathcal{K}(\mathbb{S}^n))$ of n-planes in the space of Killing tensors, which is isomorphic to the real Deligne-Mumford-Knudsen moduli space $\bar{\mathcal{M}}_{0,n+2}(\mathbb{R})$ of stable algebraic curves of genus zero with $n + 2$ marked points.*

The proof of this theorem is an application of the classification of *Gaudin subalgebras* in the *Kohno-Drinfeld Lie algebra*, a recent result of Leonardo Aguirre, Giovanni Felder and Alexander P. Veselov [AFV11].

The moduli spaces $\bar{\mathcal{M}}_{0,n}(\mathbb{R})$ have a very rich geometrical and combinatorial structure and a lot is known about their topology. We will give a brief survey in Section 3.5. By our Main Theorem III, all properties of $\bar{\mathcal{M}}_{0,n+2}(\mathbb{R})$ immediately carry over to separation coordinates and Stäckel systems on $\mathbb{S}^n$, the most important for our context being the following:

- $\bar{\mathcal{M}}_{0,n+2}(\mathbb{R})$ carries a natural free action of the permutation group S_{n+1}.
- $\bar{\mathcal{M}}_{0,n+2}(\mathbb{R})$ is tiled by $(n+1)!/2$ copies of the Stasheff polytope K_{n+1}.
- S_{n+1} acts effectively and transitively on the set of tiles.
- The stabiliser of a tile is isomorphic to $\mathbb{Z}_2$ and acts by a reflection in a hyperplane.
- $\bar{\mathcal{M}}_{0,n+2}(\mathbb{R})$ carries a natural operad structure, called the *mosaic operad*.

The final part of Chapter 3 is dedicated to the derivation of the following results from the above properties:

Corollary 3.14: A natural parametrisation of equivalence classes of separation coordinates on $\mathbb{S}^n$ by the smooth projective variety $\bar{\mathcal{M}}_{0,n+2}(\mathbb{R})$.

Section 3.6: A description of the topology of the resulting classification space of separation coordinates on $\mathbb{S}^n$:

$$\mathcal{S}(\mathbb{S}^n)/\mathrm{O}(n+1) \cong \bar{\mathcal{M}}_{0,n+2}(\mathbb{R})/S_{n+1}.$$

This solves part (iii) of Problem III for spheres of arbitrary dimension.

Section 3.7.1: A labelling of the different classes of separation coordinates on spheres in terms of the combinatorics of Stasheff polytopes.

Section 3.7.2: A description of the operad structure on orthogonal separation coordinates on spheres, resulting in an explicit construction of separation coordinates on spheres.
Section 3.7.3: The same for Stäckel systems.

In particular, we recover in this way the classical list of separation coordinates on S^n in an independent, self-contained and purely algebraic way. Moreover, since the faces of a Stasheff polytope can be labelled by rooted planar trees, our approach also provides an explanation for the graphical procedure used by Kalnins & Miller to classify separation coordinates on spheres. This finally corroborates the correctness of our approach (or, depending on your point of view, the classical results). Remarkably, the formula for the operad composition on separation coordinates already appears in [KM86], albeit without reference to operads. A uniform construction of Stäckel systems on spheres has been unknown before.

Formally, we obtain from the above construction of Stäckel systems on spheres also a parametrisation of the varieties $\mathcal{I}(S^n)$ and $\mathcal{S}(S^n)$ for arbitrary n. However, a satisfactory algebraic geometric interpretation of $\mathcal{I}(S^n)$ for $n > 3$ and of $\mathcal{S}(S^n)$ for $n > 2$ is still missing.

The results of Chapter 1 are published in [Sch12, Sch15], those of Chapter 2 can be found in [Sch14] and the results of Chapter 3 have been obtained in collaboration with Alexander P. Veselov in [SV15].

1 The foundation: the algebraic integrability conditions

Contents

In this chapter we translate the Nijenhuis integrability conditions for a Killing tensor on a constant curvature manifold into algebraic conditions on the corresponding algebraic curvature tensors. To this end, we substitute (0.7) into (0.2) and both into (0.3) and then use the representation theory for general linear groups to get rid of the dependence on the base point in the manifold.

Note that the algebraic curvature tensor in (0.7) is implicitly symmetrised in the first and third entry. The result of this operation is a tensor having the symmetries of an algebraic curvature tensor, but with antisymmetry replaced by symmetry.

Definition 1.1. *A symmetrised algebraic curvature tensor on a vector space V is an element $R \in V^* \otimes V^* \otimes V^* \otimes V^*$ satisfying the following* *symmetries:*

$$S(x, w, y, z) = +S(w, x, y, z) = S(w, x, z, y) \quad \text{(symmetry)}$$
$$S(y, z, w, x) = S(w, x, y, z) \quad \text{(pair symmetry)}$$
$$S(w, x, y, z) + S(w, y, z, x) + S(w, z, x, y) = 0 \quad \text{(Bianchi identity)}[1]$$

In subsequent computations it will be more convenient to work with the symmetrised version of algebraic curvature tensors. Actually, both representations are isomorphic.

Remark 1.2. *The space of algebraic curvature tensors on V and the space of symmetrised algebraic curvature tensors on V are isomorphic representations of* $\text{GL}(V)$. *Explicitly, this isomorphism is given by*

$$S(w, x, y, z) = \tfrac{1}{\sqrt{3}}\big(R(w, y, x, z) + R(w, z, x, y)\big) \quad (1.2a)$$
$$R(w, x, y, z) = \tfrac{1}{\sqrt{3}}\big(S(w, y, x, z) - S(w, z, x, y)\big). \quad (1.2b)$$

Since the Nijenhuis torsion of K depends on K and its covariant derivative, ∇K, we need to express both in terms of the corresponding symmetrised algebraic curvature tensor S.

Lemma 1.3. *Up to a constant factor that can be neglected, we have*

$$K_x(v, w) = S(x, x, v, w) \quad (1.3a)$$
$$(\nabla_u K)_x(v, w) = 2S(x, u, v, w). \quad (1.3b)$$

Proof. Up to said factor, the expression (1.3a) for K follows from substituting (1.2b) into (0.7). For a flat space $M \subset V$ as in (0.5b), Formula (1.3b) follows trivially. So let us assume that $M \subset V$ is as

---

[1] Owing to the other two symmetries, the cyclic sum may be taken over any three of the four entries.

in (0.5a). Denoting the covariant derivative on V by $\hat{\nabla}$, the covariant derivative of K is then given by

$$
\begin{aligned}
(\nabla_u K)&(v,w) \\
&= \nabla_u\big(K(v,w)\big) - K\big(\nabla_u v, w\big) - K\big(v, \nabla_u w\big) \\
&= \hat{\nabla}_u\big(S(x,x,v,w)\big) - S(x,x,\nabla_u v, w) - S(x,x,v,\nabla_u w) \\
&= 2S(x,\hat{\nabla}_u x, v, w) + S(x,x,\hat{\nabla}_u v, w) + S(x,x,v,\hat{\nabla}_u w) \\
&\quad - S(x,x,\hat{\nabla}_u v - g(u,v)x, w) - S(x,x,v,\hat{\nabla}_u w - g(u,w)x) \\
&= 2S(x,u,v,w).
\end{aligned}
$$

For the last equality we used the fact that the Bianchi identity for S implies that $S(x,x,x,w) = 0$ and $S(x,x,v,x) = 0$. $\qquad\square$

The proof of Proposition 0.9 is now a simple consequence of the above lemma, so we will give it here for the sake of completeness.

Proof (of Proposition 0.9). We have to show that the map defined by (1.3a) is an isomorphism between Killing tensors on $M \subset V$ and symmetrised algebraic curvature tensors on V. This map is well defined, since by (1.3b) the Killing equation for (1.3a) is equivalent to the Bianchi identity for S. For simplicity let us assume that $M \subset V$ is not flat, i.e. of the form (0.5a). To show the injectivity of the above map, suppose $S(x,x,v,w) = 0$ for all $x,v,w \in V$ with $g(x,x) = 1$ and $g(v,x) = g(w,x) = 0$. We can omit the restriction $g(v,x) = g(w,x) = 0$ due to the Bianchi identity for S. We can also omit the restriction $g(x,x) = 1$, because $S(x,x,v,w)$ is a homogeneous polynomial in x for fixed $v,w \in V$ and $\mathbb{R}M$ is open in V. From a polarisation in x we then conclude that $S = 0$. The surjectivity of the above map now follows from dimension considerations. Indeed, the dimension of the space of Killing tensors on a constant curvature manifold of dimension n is known to be

$$
\frac{(n+1)n^2(n-1)}{12},
$$

which happens to be the dimension of the space of algebraic curvature tensors in dimension $n + 1$.[2] For a flat space $M \subset V$ as in (0.5b) the proof is analogous and will be left to the reader.   □

For actual computations the use of index notation is indispensable. We will write Greek indices $\alpha, \beta, \gamma, \ldots$ for local coordinates on M (ranging from 1 to n) and Latin indices $a, b, c, \ldots$ for components in V (ranging from 0 to n). We can then denote both, the inner product on V as well as the induced metric on M, by the same letter g and distinguish them only via the type of indices. Consequently, Latin indices are raised and lowered using g_{ab} and greek ones using $g_{\alpha\beta}$.

This said, we can rewrite the expressions (1.3) using $\nabla_v x^a = v^a$ as

$$K_{\alpha\beta} = S_{a_1 a_2 b_1 b_2} x^{a_1} x^{a_2} \nabla_\alpha x^{b_1} \nabla_\beta x^{b_2} \tag{1.4a}$$

$$\nabla_\gamma K_{\alpha\beta} = 2 S_{c_1 c_2 d_1 d_2} x^{c_1} \nabla_\gamma x^{c_2} \nabla_\alpha x^{d_1} \nabla_\beta x^{d_2}, \tag{1.4b}$$

where we regard the components x^a of $x \in V$ as functions on $M \subset V$ by restriction. We are now ready to substitute (1.4) into (0.2) and then further into (0.3).

First note that in the integrability conditions (0.3) the Nijenhuis torsion (0.2) appears only antisymmetrised in its two lower indices β and γ. To simplify computations we will thus replace the Nijenhuis torsion $N^\alpha{}_{\beta\gamma}$ in the integrability conditions by the tensor

$$\bar{N}^\alpha{}_{\beta\gamma} := \tfrac{1}{2}\left(K^\alpha{}_\delta \nabla_\gamma K^\delta{}_\beta + K^\delta{}_\beta \nabla_\delta K^\alpha{}_\gamma\right), \qquad \bar{N}^\alpha{}_{[\beta\gamma]} = N^\alpha{}_{\beta\gamma}.$$

Together with (1.4) this can be written as

$$\bar{N}^\alpha{}_{\beta\gamma} = S_{a_1 a_2 b_1 b_2} S_{c_1 c_2 d_1 d_2} x^{a_1} x^{a_2} x^{c_1} \nabla^\alpha x^{b_1} \nabla_\delta x^{b_2} \nabla_\gamma x^{c_2} \nabla^\delta x^{d_1} \nabla_\beta x^{d_2}$$
$$+ S_{a_1 a_2 b_1 b_2} S_{c_1 c_2 d_1 d_2} x^{a_1} x^{a_2} x^{c_1} \nabla^\delta x^{b_1} \nabla_\beta x^{b_2} \nabla_\delta x^{c_2} \nabla^\alpha x^{d_1} \nabla_\gamma x^{d_2}.$$

Lemma 1.4. *For a constant curvature manifold we have*

$$\nabla_\delta x^a \nabla^\delta x^b = \begin{cases} g^{ab} - x^a x^b & \text{if } M \text{ is of the form (0.5a)} \\ g^{ab} - u^a u^b & \text{if } M \text{ is of the form (0.5b)}. \end{cases}$$

---

[2]This can be computed from the so called *hook formula*.

Proof. Let $e_1, \ldots, e_n$ be a basis of $T_x M$ and complete it with a unit normal vector $u =: e_0$ to a basis of V. Then on one hand

$$
\sum_{i,j=0}^{n} g(e^i, e^j) \nabla_{e_i} x^a \nabla_{e_j} x^b = \sum_{i,j=1}^{n} g(e^i, e^j) \nabla_{e_i} x^a \nabla_{e_j} x^b + \nabla_u x^a \nabla_u x^b
$$
$$
= g^{\alpha\beta} \nabla_\alpha x^a \nabla_\beta x^b + u^a u^b.
$$

On the other hand, choosing the standard basis of V instead, the left hand side is just g^{ab}. This proves the lemma, remarking that $u = x$ if M is not flat. $\qquad\square$

For flat M the lemma yields

$$
\begin{aligned}
\bar{N}^\alpha{}_{\beta\gamma} &= \bar{g}^{b_2 d_1} S_{a_1 a_2 b_1 b_2} S_{c_1 c_2 d_1 d_2} x^{a_1} x^{a_2} x^{c_1} \nabla^\alpha x^{b_1} \nabla_\beta x^{d_2} \nabla_\gamma x^{c_2} \\
&+ \bar{g}^{b_1 c_2} S_{a_1 a_2 b_1 b_2} S_{c_1 c_2 d_1 d_2} x^{a_1} x^{a_2} x^{c_1} \nabla^\alpha x^{d_1} \nabla_\beta x^{b_2} \nabla_\gamma x^{d_2},
\end{aligned}
\tag{1.5}
$$

where $\bar{g} := g^{ab} - u^a u^b$. In all other cases we have

$$
\begin{aligned}
\bar{N}^\alpha{}_{\beta\gamma} &\\
&= \left(g^{b_2 d_1} - x^{b_2} x^{d_1} \right) S_{a_1 a_2 b_1 b_2} S_{c_1 c_2 d_1 d_2} x^{a_1} x^{a_2} x^{c_1} \nabla^\alpha x^{b_1} \nabla_\beta x^{d_2} \nabla_\gamma x^{c_2} \\
&+ \left(g^{b_1 c_2} - x^{b_1} x^{c_2} \right) S_{a_1 a_2 b_1 b_2} S_{c_1 c_2 d_1 d_2} x^{a_1} x^{a_2} x^{c_1} \nabla^\alpha x^{d_1} \nabla_\beta x^{b_2} \nabla_\gamma x^{d_2}.
\end{aligned}
$$

But here the two subtracted terms vanish by the Bianchi identity, because they contain the terms

$$
S_{a_1 a_2 b_1 b_2} x^{a_1} x^{a_2} x^{b_2} = 0, \qquad S_{a_1 a_2 b_1 b_2} x^{a_1} x^{a_2} x^{b_1} = 0.
$$

This allows us to use (1.5) for *all* constant curvature manifolds of the form (0.5) if we define

$$
\bar{g}^{ab} := \begin{cases} g^{ab} & \text{if } M \text{ is of the form (0.5a)} \\ g^{ab} - u^a u^b & \text{if } M \text{ is of the form (0.5b)}. \end{cases}
\tag{1.6}
$$

In the case of a hyperplane $M \subset V$, the tensor $\bar{g}^{ab}$ is the pullback of the metric on M via the orthogonal projection $V \to M$ and thus degenerated. Note that we still lower and rise indices with the metric g^{ab} and not with $\bar{g}^{ab}$.

In (1.5) the lower indices b_2, d_1 respectively b_1, c_2 are contracted with $\bar{g}$. We can make use of the symmetries of $S_{a_1 a_2 b_1 b_2}$ to bring these indices to the first position:

$$\bar{N}^\alpha{}_{\beta\gamma} = \bar{g}^{b_2 d_1} S_{b_2 b_1 a_1 a_2} S_{d_1 d_2 c_1 c_2} x^{a_1} x^{a_2} x^{c_1} \nabla^\alpha x^{b_1} \nabla_\beta x^{d_2} \nabla_\gamma x^{c_2}$$
$$+ \bar{g}^{b_1 c_2} S_{b_1 b_2 a_1 a_2} S_{c_2 c_1 d_1 d_2} x^{a_1} x^{a_2} x^{c_1} \nabla^\alpha x^{d_1} \nabla_\beta x^{b_2} \nabla_\gamma x^{d_2} .$$

Renaming, lowering and rising indices appropriately finally results in

$$\bar{N}_{\alpha\beta\gamma} = \bar{g}_{ij} \left(S^i{}_{a_2 b_1 b_2} S^j{}_{c_2 d_1 d_2} + S^i{}_{c_2 b_1 b_2} S^j{}_{d_1 a_2 d_2} \right)$$
$$x^{b_1} x^{b_2} x^{d_1} \nabla_\alpha x^{a_2} \nabla_\beta x^{c_2} \nabla_\gamma x^{d_2} . \quad (1.7)$$

In what follows we will substitute this expression together with (1.4a) into each of the three integrability conditions (0.3) and transform them into purely algebraic integrability conditions.

1.1 Young tableaux

Throughout this chapter we will use Young tableaux as a compact means for index manipulations on tensors with many indices. The reader not familiar with this formalism may as well simply consider them as an alternative notation for symmetrisation and antisymmetrisation operators. However, we prefer Young tableaux over the more common notation using round respectively square brackets around the indices to be symmetrised, as the latter becomes confusing when several index sets are involved and even ambiguous if these sets are not disjoint. Moreover, using Young tableaux has the additional advantage that one can directly read off the symmetry class of the tensors involved. As we will basically deal with only a single type of Young tableaux, namely those of a "hook shape", we introduce them by means of examples. More details can be found in [Sch12]. For the background we refer the reader to the standard literature on representation theory of symmetric and linear groups.

Young tableaux define elements in the group algebra of the permutation group S_d. That is, a Young tableau stands for a (formal) linear combination of permutations of d objects. In our case, these objects will be certain tensor indices. For the sake of simplicity of notation we will identify a Young tableau with the group algebra element it defines. A Young tableau consisting of a single row denotes the sum of all permutations of the indices in this row. For example, using cycle notation,

$$\boxed{a_2\ c_1\ c_2} = e + (a_2 c_1) + (c_1 c_2) + (c_2 a_2) + (a_2 c_1 c_2) + (c_2 c_1 a_2).$$

This is an element in the group algebra of the group of permutations of the indices a_2, c_1 and c_2 (or any superset). In the same way a Young tableau consisting of a single column denotes the *signed* sum of all permutations of the indices in this column, the sign being the sign of the permutation. For example,

$$\boxed{\begin{array}{c} a_1 \\ b_1 \\ d_2 \end{array}} = e - (a_1 b_1) - (b_1 d_2) - (d_2 a_1) + (a_1 b_1 d_2) + (d_2 b_1 a_1).$$

We call these *row symmetrisers* respectively *column antisymmetrisers*. The reason we define them without the usual normalisation factors is that then all numerical constants appear explicitly in our computations (although irrelevant for our concerns).

The group multiplication extends linearly to a natural product in the group algebra. A general Young tableau is then simply the product of all row symmetrisers and all column antisymmetrisers of the tableau. We will only deal with Young tableaux having a "hook shape", such as the following:

$$\boxed{\begin{array}{cccc} a_1 & a_2 & c_1 & c_2 \end{array}}_{\substack{b_1 \\ d_2}} = \boxed{a_1\ a_2\ c_1\ c_2}\ \boxed{\begin{array}{c} a_1 \\ b_1 \\ d_2 \end{array}}. \tag{1.8a}$$

The inversion of group elements extends linearly to an involution of the group algebra. If we consider elements in the group algebra as linear operators on the group algebra itself, this involution is the adjoint with respect to the natural inner product on the group algebra, given by defining the group elements to be an orthonormal basis. Since this operation leaves symmetrisers and antisymmetrisers invariant, it simply exchanges the order of symmetrisers and antisymmetrisers in a Young tableau. The adjoint of (1.8a) for example is

$$
\left[\begin{array}{c}\boxed{a_1\,a_2\,c_1\,c_2}\\\boxed{b_1}\\\boxed{d_2}\end{array}\right]^{\star} = \begin{array}{c}\boxed{a_1}\\\boxed{b_1}\\\boxed{d_2}\end{array}\ \boxed{a_1\,a_2\,c_1\,c_2}\,. \tag{1.8b}
$$

Properly scaled, Young tableaux with d boxes define projectors onto irreducible S_d-representations. A hook shaped Young tableau with p rows and q columns for example satisfies

$$
\left[\begin{array}{c}\boxed{a\,b\,\cdots\,c}\\\boxed{d}\\\vdots\\\boxed{e}\end{array}\right]^{2} = (p+q-1)(p-1)!(q-1)!\ \begin{array}{c}\boxed{a\,b\,\cdots\,c}\\\boxed{d}\\\vdots\\\boxed{e}\end{array} \tag{1.9}
$$

and the same formula holds for its adjoint.

The isomorphism class of the irreducible representation defined by a Young tableau is labelled by the corresponding *Young frame*, which is the Young tableau with the labels of its boxes erased. On the level of isomorphism classes, the decomposition of tensor products of irreducible representations is given by the *Littlewood-Richardson rule*. For example, according to this rule, the tensor product of a symmetric and an antisymmetric representation decomposes into two irreducible components, each of hook symmetry:

$$
q\left\{\ \begin{array}{c}\vdots\end{array}\ \otimes\ \overbrace{\boxed{\cdots}}^{p}\ \cong\ \overbrace{\begin{array}{c}\boxed{\ \ \cdots\ \ }\\\vdots\\\ \end{array}}^{p}\ \oplus\ \begin{array}{c}\boxed{\ \ \cdots\ \ }\\\vdots\\\ \end{array}\right\}q\,. \tag{1.10}
$$

The following lemma gives an explicit realisation of this decomposition in terms of orthogonal projectors.

Lemma 1.5.

$$\frac{1}{q!}\boxed{\begin{smallmatrix}a_1\\ \vdots\\ a_q\end{smallmatrix}} \cdot \frac{1}{p!}\boxed{s_1 \cdots s_p} = \frac{\frac{p}{q+1}}{(p+q)p!^2 q!^2}\boxed{\begin{smallmatrix}s_1 \cdots s_p\\ a_1\\ \vdots\\ a_q\end{smallmatrix}}\,\boxed{\begin{smallmatrix}s_1 \cdots s_p\\ a_1\\ \vdots\\ a_q\end{smallmatrix}}^{\!\star}$$

$$+ \frac{\frac{q}{p+1}}{(p+q)p!^2 q!^2}\boxed{\begin{smallmatrix}a_1 s_1 \cdots s_p\\ \vdots\\ a_q\end{smallmatrix}}^{\!\star}\,\boxed{\begin{smallmatrix}a_1 s_1 \cdots s_p\\ \vdots\\ a_q\end{smallmatrix}} \tag{1.11}$$

In particular, for $p = q = 3$:

$$\frac{1}{3!}\boxed{\begin{smallmatrix}c_2\\ d_2\\ a_2\end{smallmatrix}} \cdot \frac{1}{3!}\boxed{b_2\,b_1\,d_1} = \frac{1}{2^7 3^4}\boxed{\begin{smallmatrix}b_2\,b_1\,d_1\\ c_2\\ d_2\\ a_2\end{smallmatrix}}\,\boxed{\begin{smallmatrix}b_2\,b_1\,d_1\\ c_2\\ d_2\\ a_2\end{smallmatrix}}^{\!\star} + \frac{1}{2^7 3^4}\boxed{\begin{smallmatrix}c_2\,b_2\,b_1\,d_1\\ d_2\\ a_2\end{smallmatrix}}^{\!\star}\,\boxed{\begin{smallmatrix}c_2\,b_2\,b_1\,d_1\\ d_2\\ a_2\end{smallmatrix}}.$$

$$\tag{1.12}$$

Proof. Write (1.11) as $P = P_1 + P_2$. Decomposing temporarily the hook symmetrisers on the right hand side as in (1.8) into a product of a symmetriser and an antisymmetriser and using (1.9), one easily checks that P, P_1 and P_2 are orthogonal projectors verifying $P_1 P_2 = 0 = P_2 P_1$, $P P_1 = P_1$ and $P P_2 = P_2$. Therefore $P_1 + P_2$ is an orthogonal projector with image $\operatorname{im} P_1 \oplus \operatorname{im} P_2 \subseteq \operatorname{im} P$. The decomposition of the isomorphism class of $\operatorname{im} P$ into irreducible components is given by (1.10). The Young frames on the right hand side are those appearing in the expression for P_1 respectively P_2. Hence they describe the isomorphism classes of $\operatorname{im} P_1$ and $\operatorname{im} P_2$. This shows that $\operatorname{im} P$ and $\operatorname{im}(P_1 + P_2) = \operatorname{im} P_1 \oplus \operatorname{im} P_2$ have the same dimension and are thus equal. This implies $P = P_1 + P_2$. $\qquad\square$

Remark 1.6. *The lemma can be interpreted as an explicit splitting of the terms in the long exact sequence*

$$0 \to \Lambda^d V \to \ldots \to S^p V \otimes \Lambda^q V \to S^{p+1} V \otimes \Lambda^{q-1} V \to \ldots \to S^d V \to 0,$$

known as the Koszul complex.

The permutation group S_d acts on d-fold covariant or contravariant tensors by permuting indices. This action extends linearly to an action of the entire group algebra. In particular, any Young tableau acts on tensors with corresponding indices. For example,

$$\boxed{\begin{smallmatrix} b_1 \\ a_2 \\ c_2 \end{smallmatrix}}\, T_{b_1 a_2 c_2} = T_{b_1 a_2 c_2} - T_{a_2 b_1 c_2} - T_{b_1 c_2 a_2} - T_{c_1 a_2 b_2} + T_{a_1 c_2 b_2} + T_{c_1 b_2 a_2}.$$

To give another example, the operator (1.8a) acts on a tensor

$$T_{b_1 b_2 d_1 d_2 a_2 c_2}$$

by an antisymmetrisation in the indices b_1, a_2, c_2 and a subsequent symmetrisation in the indices b_1, b_2, d_1, d_2. In the same way its adjoint (1.8b) acts by first symmetrising and then antisymmetrising.

1.2 The 1st integrability condition

The first integrability condition (0.3a) can be written as $\bar{N}_{[\alpha\beta\gamma]} = 0$. For the expression (1.7) this is equivalent to the vanishing of the antisymmetrisation in the upper indices a_2, c_2, d_2:

$$\bar{g}_{ij}\left(S^i{}_{a_2 b_1 b_2} S^j{}_{c_2 d_1 d_2} + S^i{}_{c_2 b_1 b_2} S^j{}_{d_1 a_2 d_2}\right)$$
$$x^{b_1} x^{b_2} x^{d_1} \nabla_\alpha x^{[a_2} \nabla_\beta x^{c_2} \nabla_\gamma x^{d_2]} = 0.$$

Due to the symmetry of $S^j{}_{d_1 a_2 d_2}$ in a_2, d_2 the second term vanishes. If we write u, v and w for the tangent vectors ∂_α, ∂_β respectively ∂_γ and use $\nabla_u x^a = u^a$ in order to get rid of the ∇'s, we obtain the condition

$$\bar{g}_{ij} S^i{}_{a_2 b_1 b_2} S^j{}_{c_2 d_1 d_2} x^{b_1} x^{b_2} x^{d_1} u^{[a_2} v^{c_2} w^{d_2]} = 0$$

$$\forall x \in M, \quad \forall u, v, w \in T_x M$$

(1.13)

on the symmetrised algebraic curvature tensor S.

Applying (1.12) to the tensor $x^{b_1}x^{b_2}x^{d_1}u^{[a_2}v^{c_2}w^{d_2]}$ and decomposing the hook symmetrisers as in (1.8), we conclude that

$$x^{b_1}x^{b_2}x^{d_1}u^{[a_2}v^{c_2}w^{d_2]}$$

$$= \frac{1}{2^7 3^4}\left(\boxed{\begin{array}{ccc} b_2 & b_1 & d_1 \end{array}}\!\!\begin{array}{c} \boxed{b_2} \\ \boxed{c_2} \\ \boxed{d_2} \\ \boxed{a_2} \end{array}^{2} + \boxed{\begin{array}{ccc} b_2 & b_1 & d_1 \end{array}}\,\boxed{\begin{array}{cccc} c_2 & b_2 & b_1 & d_1 \end{array}}\begin{array}{c}\boxed{c_2}\\\boxed{d_2}\\\boxed{a_2}\end{array}^{2}\boxed{\begin{array}{c}c_2\\d_2\\a_2\end{array}} \right)$$

$$x^{b_1}x^{b_2}x^{d_1}u^{[a_2}v^{c_2}w^{d_2]}$$

$$= \frac{1}{2^3 3^2}\left(\begin{array}{c}\boxed{\begin{array}{ccc}b_2&b_1&d_1\end{array}}\\\boxed{c_2}\\\boxed{d_2}\\\boxed{a_2}\end{array} + \begin{array}{c}\boxed{\begin{array}{cccc}c_2&b_2&b_1&d_1\end{array}}\\\boxed{d_2}\\\boxed{a_2}\end{array}^{\star} \right) x^{b_1}x^{b_2}x^{d_1}u^{[a_2}v^{c_2}w^{d_2]}.$$

Substituted into (1.13), we get

$$\bar{g}_{ij}S^i{}_{a_2b_1b_2}S^j{}_{c_2d_1d_2}$$

$$\left(\left(\begin{array}{c}\boxed{\begin{array}{ccc}b_2&b_1&d_1\end{array}}\\\boxed{c_2}\\\boxed{d_2}\\\boxed{a_2}\end{array} + \begin{array}{c}\boxed{\begin{array}{cccc}c_2&b_2&b_1&d_1\end{array}}\\\boxed{d_2}\\\boxed{a_2}\end{array}^{\star} \right) x^{b_1}x^{b_2}x^{d_1}u^{[a_2}v^{c_2}w^{d_2]} \right) = 0. \quad (1.14)$$

Symmetrising the upper indices in a contraction is equivalent to symmetrising the corresponding lower indices and the same is true for antisymmetrisation. Hence we can replace a Young tableau acting on upper indices in a contraction by its dual acting on the corresponding lower indices. Consequently, the previous equation is equivalent to

$$\left(\left(\begin{array}{c}\boxed{\begin{array}{ccc}b_2&b_1&d_1\end{array}}\\\boxed{c_2}\\\boxed{d_2}\\\boxed{a_2}\end{array}^{\star} + \begin{array}{c}\boxed{\begin{array}{cccc}c_2&b_2&b_1&d_1\end{array}}\\\boxed{d_2}\\\boxed{a_2}\end{array} \right) \bar{g}_{ij}S^i{}_{a_2b_1b_2}S^j{}_{c_2d_1d_2} \right)$$

$$x^{b_1}x^{b_2}x^{d_1}u^{[a_2}v^{c_2}w^{d_2]} = 0. \quad (1.15)$$

The following lemma shows that the second of the two terms in (1.15) and hence also in (1.14) vanishes identically.

Lemma 1.7.

$$\begin{array}{c}\boxed{\begin{array}{cccc}c_2&b_2&b_1&d_1\end{array}}\\\boxed{d_2}\\\boxed{a_2}\end{array}\ \bar{g}_{ij}S^i{}_{a_2b_1b_2}S^j{}_{c_2d_1d_2} = 0 \quad (1.16)$$

Before we prove the lemma, we mention an identity which we will frequently use and which is obtained from symmetrising the Bianchi identity

$$S^i{}_{a_1 b_1 b_2} + S^i{}_{b_1 b_2 a_1} + S^i{}_{b_2 a_1 b_1} = 0 \tag{1.17}$$

in b_1, b_2:

$$\boxed{b_1\,b_2}\, S^i{}_{a_2 b_1 b_2} = -2\,\boxed{b_1\,b_2}\, S^i{}_{b_1 b_2 a_2}. \tag{1.18}$$

We refer to this identity as *symmetrised Bianchi identity*.

Proof. The left hand side of (1.16) is

$$\boxed{\begin{smallmatrix} c_2 & b_2 & b_1 & d_1 \\ d_2 \\ a_2 \end{smallmatrix}}\, \bar{g}_{ij} S^i{}_{a_2 b_1 b_2} S^j{}_{c_2 d_1 d_2} = \boxed{\begin{smallmatrix} c_2 & b_2 & b_1 & d_1 \\ d_2 \\ a_2 \end{smallmatrix}}\, \bar{g}_{ij} S^i{}_{a_2 b_1 b_2} S^j{}_{c_2 d_1 d_2}$$

$$= \boxed{c_2\,b_2\,b_1\,d_1}\, g_{ij} \big(S^i{}_{a_2 b_1 b_2} S^j{}_{c_2 d_1 d_2} - S^i{}_{d_2 b_1 b_2} S^j{}_{c_2 d_1 a_2}$$
$$+ S^i{}_{d_2 b_1 b_2} S^j{}_{a_2 d_1 c_2} - S^i{}_{a_2 b_1 b_2} S^j{}_{d_2 d_1 c_2}$$
$$+ S^i{}_{c_2 b_1 b_2} S^j{}_{d_2 d_1 a_2} - S^i{}_{c_2 b_1 b_2} S^j{}_{a_2 d_1 d_2} \big)$$

Regard the parenthesis under complete symmetrisation in c_2, b_2, b_1 and d_1. The last two terms vanish due to the Bianchi identity (1.17). Renaming i, j as j, i in the third term shows that it cancels the fourth. That the first two also cancel each other can be seen by applying twice the symmetrised Bianchi identity (1.18), once to $S^i{}_{a_2 b_1 b_2}$ and once to $S^i{}_{c_2 d_1 d_2}$. $\qquad\square$

Remark 1.8. *Abstractly, the above lemma also follows from the symmetry classification of Riemann tensor polynomials [FKWC92], since the tensor*

$$g_{ij} R^i{}_{a_2 b_1 b_2} R^j{}_{c_2 d_1 d_2}$$

has symmetry type

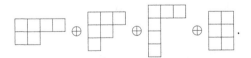

Resuming, the first integrability condition is equivalent to

$$\bar{g}_{ij} S^i{}_{a_2 b_1 b_2} S^j{}_{c_2 d_1 d_2} \left(\young(b_2b_1d_1,c_2,d_2,a_2)\ x^{b_1} x^{b_2} x^{d_1} u^{a_2} v^{c_2} w^{d_2} \right) = 0 \tag{1.19}$$

$$\forall x \in M, \quad \forall u, v, w \in T_x M.$$

Here we omitted the explicit antisymmetrisation in a_2, c_2, d_2, since this is already carried out implicitly by the Young symmetriser. We can drop the restriction $u, v, w \in T_x M$ in (1.19). Indeed, we can decompose $u, v, w \in V$ according to the splitting $V = T_x M \oplus \mathbb{R}x$ and Dirichlet's drawer principle shows that

$$\young(b_2b_1d_1,c_2,d_2,a_2)\ x^{b_1} x^{b_2} x^{d_1} u^{a_2} v^{c_2} w^{d_2} = 0 \qquad \text{if } u = x \text{ or } v = x \text{ or } w = x.$$

This trick is crucial, as it allows us to deal with Young projectors instead of the much more complicated projectors onto $O(V)$-representations. We can also drop the restriction $x \in M$, since (1.19) is a homogeneous polynomial in x for fixed $u, v, w \in V$ and $\mathbb{R}M$ is open in V.

As before, we can replace the Young symmetriser in (1.19) acting on upper indices by its adjoint acting on lower indices. Condition (1.19) is therefore equivalent to

$$\left(\young(b_2b_1d_1,c_2,d_2,a_2)^{\star}\ \bar{g}_{ij} S^i{}_{a_2 b_1 b_2} S^j{}_{c_2 d_1 d_2} \right) x^{b_1} x^{b_2} x^{d_1} u^{a_2} v^{c_2} w^{d_2} = 0$$

without any restriction on $x, u, v, w \in V$. Polarisation in x now yields the first integrability condition in the purely algebraic form[3]

$$\young(b_2b_1d_1,c_2,d_2,a_2)^{\star}\ \bar{g}_{ij} S^i{}_{a_2 b_1 b_2} S^j{}_{c_2 d_1 d_2} = 0. \tag{1.20}$$

---

[3] At this point we reduce the infinitely many algebraic equations parametrised by $x \in M$ and $u, v, w \in T_x M$ to a finite set, as predicted by Hilbert's basis theorem.

We will now give a number of equivalent formulations.

Proposition 1.9. *Any of the following conditions is equivalent to the first Nijenhuis integrability condition* (0.3a) *for a Killing tensor on a constant curvature manifold* M.

(i) The corresponding symmetrised algebraic curvature tensor S *satisfies*

$$P \, \bar{g}_{ij} S^i{}_{a_2 b_1 b_2} S^j{}_{c_2 d_1 d_2} = 0, \tag{1.21}$$

where P *is any of the following pseudo-projectors.*

$$(a) \;\; \boxed{\begin{smallmatrix} b_2 & b_1 & d_1 \\ c_2 \\ d_2 \\ a_2 \end{smallmatrix}}^{\!\star} \qquad (b) \;\; \boxed{\begin{smallmatrix} c_2 \\ d_2 \\ a_2 \end{smallmatrix}}\,\boxed{b_2\,b_1\,d_1} \qquad (c) \;\; \boxed{\begin{smallmatrix} b_2 \\ c_2 \\ d_2 \\ a_2 \end{smallmatrix}} \qquad (d) \;\; \boxed{\begin{smallmatrix} b_2 & b_1 & d_1 \\ c_2 \\ d_2 \\ a_2 \end{smallmatrix}} \tag{1.22}$$

(ii) The corresponding algebraic curvature tensor R *satisfies*

$$P \, \bar{g}_{ij} R^i{}_{b_1 a_2 b_2} R^j{}_{d_1 c_2 d_2} = 0, \tag{1.23}$$

where P *is any of the symmetry operators* (1.22).

If M *is not flat, this is is in addition equivalent to:*

(iii) The curvature form $\Omega \in \mathrm{End}(V) \otimes \Lambda^2 V$ *of* R, *defined by*

$$\Omega^{a_1}{}_{b_1} := R^{a_1}{}_{b_1 a_2 b_2} dx^{a_2} \wedge dx^{b_2}$$

satisfies

$$\Omega \wedge \Omega = 0, \tag{1.24}$$

where the wedge product is defined by taking the exterior product in the $\Lambda^2 V$-*component and usual matrix multiplication in the* $\mathrm{End}(V)$-*component.*

Proof. We have already shown that the first integrability condition (0.3a) is equivalent to (1.20). This is condition (1.21a). The equivalence (1.21a) $\Leftrightarrow$ (1.21b) follows from (1.12) combined with (1.16).

The implication (1.21c) $\Rightarrow$ (1.21d) is trivial. We finish the proof of part (i) by proving (1.21d) $\Rightarrow$ (1.21b) $\Rightarrow$ (1.21c) through a stepwise manipulation of

$$
\boxed{\begin{array}{ccc} b_2 & b_1 & d_1 \\ \hline c_2 \\ d_2 \\ a_2 \end{array}}\; \bar{g}_{ij} S^i{}_{a_2 b_1 b_2} S^j{}_{c_2 d_1 d_2} = \boxed{\begin{array}{ccc} & b_2 & \\ \hline b_2 & b_1 & d_1 \\ c_2 \\ d_2 \\ a_2 \end{array}}\; \bar{g}_{ij} S^i{}_{a_2 b_1 b_2} S^j{}_{c_2 d_1 d_2}. \tag{1.25a}
$$

In order to sum over all $q!$ permutations of q indices, one can take the sum over q cyclic permutations, choose one index and then sum over all $(q-1)!$ permutations of the remaining $(q-1)$ indices. Apply this to the antisymmetrisation in a_2, b_2, c_2, d_2 (fixing b_2):

$$
(1.25a) = \boxed{\begin{array}{ccc} b_2 & b_1 & d_1 \end{array}}\boxed{\begin{array}{c} c_2 \\ d_2 \\ a_2 \end{array}} \bar{g}_{ij} \Big(S^i{}_{\underline{a_2} b_1 b_2} S^j{}_{\underline{c_2} d_1 \underline{d_2}} - S^i{}_{b_2 b_1 \underline{c_2}} S^j{}_{\underline{d_2} d_1 \underline{a_2}}
$$
$$
+ S^i{}_{\underline{c_2} b_1 \underline{d_2}} S^j{}_{\underline{a_2} d_1 b_2} - S^i{}_{\underline{d_2} b_1 \underline{a_2}} S^j{}_{b_2 d_1 \underline{c_2}} \Big). \tag{1.25b}
$$

For a better readability we underlined each antisymmetrised index. Now use the symmetrised Bianchi identity (1.18) to bring the index c_2 from the fourth to the second index position:

$$
(1.25b) = \boxed{\begin{array}{ccc} b_2 & b_1 & d_1 \end{array}}\boxed{\begin{array}{c} c_2 \\ d_2 \\ a_2 \end{array}} \bar{g}_{ij} \Big(S^i{}_{\underline{a_2} b_1 b_2} S^j{}_{\underline{c_2} d_1 \underline{d_2}} + \tfrac{1}{2} S^i{}_{\underline{c_2} b_1 b_2} S^j{}_{\underline{d_2} d_1 \underline{a_2}}
$$
$$
+ S^i{}_{\underline{c_2} b_1 \underline{d_2}} S^j{}_{\underline{a_2} d_1 b_2} + \tfrac{1}{2} S^i{}_{\underline{d_2} b_1 \underline{a_2}} S^j{}_{\underline{c_2} d_1 b_2} \Big). \tag{1.25c}
$$

Then rename i, j as j, i in the last two terms:

$$
(1.25c) = \boxed{\begin{array}{ccc} b_2 & b_1 & d_1 \end{array}}\boxed{\begin{array}{c} c_2 \\ d_2 \\ a_2 \end{array}} \bar{g}_{ij} \Big(S^i{}_{\underline{a_2} b_1 b_2} S^j{}_{\underline{c_2} d_1 \underline{d_2}} + \tfrac{1}{2} S^i{}_{\underline{c_2} b_1 b_2} S^j{}_{\underline{d_2} d_1 \underline{a_2}}
$$
$$
+ S^i{}_{\underline{a_2} d_1 b_2} S^j{}_{\underline{c_2} b_1 \underline{d_2}} + \tfrac{1}{2} S^i{}_{\underline{c_2} d_1 b_2} S^j{}_{\underline{d_2} b_1 \underline{a_2}} \Big). \tag{1.25d}
$$

Finally use the symmetrisation in b_2, b_1, d_1 and the antisymmetrisation in c_2, d_2, a_2 to bring each term to the same form:

$$
(1.25d) = \boxed{\begin{array}{ccc} b_2 & b_1 & d_1 \end{array}}\boxed{\begin{array}{c} c_2 \\ d_2 \\ a_2 \end{array}} \bar{g}_{ij} \Big(3 S^i{}_{\underline{a_2} b_1 b_2} S^j{}_{\underline{c_2} d_1 \underline{d_2}} \Big). \tag{1.25e}
$$

This proves (1.21d) $\Leftrightarrow$ (1.21b). To continue, antisymmetrise

$$0 = \boxed{\begin{array}{c} c_2 \\ d_2 \\ a_2 \end{array}} \boxed{b_2 | b_1 | d_1} \, \bar{g}_{ij} S^i{}_{\underline{a}_2 b_1 b_2} S^j{}_{\underline{c}_2 d_1 \underline{d}_2}$$

$$= 2 \boxed{\begin{array}{c} c_2 \\ d_2 \\ a_2 \end{array}} \bar{g}_{ij} \left(S^i{}_{\underline{a}_2 b_1 b_2} S^j{}_{\underline{c}_2 d_1 \underline{d}_2} + S^i{}_{\underline{a}_2 b_2 d_1} S^j{}_{\underline{c}_2 b_1 \underline{d}_2} + S^i{}_{\underline{a}_2 d_1 b_1} S^j{}_{\underline{c}_2 b_2 \underline{d}_2} \right)$$

in a_2, b_2, c_2, d_2. Then the last term vanishes by the symmetry of $S^j{}_{c_2 b_2 d_2}$ in b_2, d_2 and yields

$$0 = \boxed{\begin{array}{c} a_2 \\ b_2 \\ c_2 \\ d_2 \end{array}} \bar{g}_{ij} \left(S^i{}_{\underline{a}_2 b_1 \underline{b}_2} S^j{}_{\underline{c}_2 d_1 \underline{d}_2} + S^i{}_{\underline{a}_2 d_1 \underline{b}_2} S^j{}_{\underline{c}_2 b_1 \underline{d}_2} \right).$$

Both sum terms are equal under antisymmetrisation in a_2, b_2, c_2, d_2 and contraction with $\bar{g}_{ij}$. Indeed, exchanging b_1 and d_1 is tantamount to exchanging a_2 with c_2 and b_2 with d_2 and renaming i, j as j, i. This proves (1.21b) $\Rightarrow$ (1.21c).

From the correspondence (1.2) between R and S we conclude the equivalence (1.21c) $\Leftrightarrow$ (1.23c). The proof of the remaining part of (ii) is completely analogous to the proof of (i), so we leave it to the reader. Condition (1.24) is just a reformulation of (1.23c). This finishes the proof. $\qquad\qquad\qquad\qquad\qquad\qquad\qquad\qquad\qquad\qquad\qquad\qquad\quad\square$

1.3 The 2nd integrability condition

The proceeding for the second integrability condition is similar. We begin by substituting the expressions (1.7) and (1.4a) into the tensor appearing in (0.3b):

$$\bar{N}^\delta{}_{\beta\gamma} K_{\delta\alpha} = \bar{g}_{ij} \left(S^i{}_{a_2 b_1 b_2} S^j{}_{c_2 d_1 d_2} + S^i{}_{c_2 b_1 b_2} S^j{}_{d_1 a_2 d_2} \right)$$
$$x^{b_1} x^{b_2} x^{d_1} \nabla^\delta x^{a_2} \nabla_\beta x^{c_2} \nabla_\gamma x^{d_2}$$
$$S_{e_1 e_2 f_1 f_2} x^{e_1} x^{e_2} \nabla_\delta x^{f_1} \nabla_\alpha x^{f_2}.$$

As before, we replace the contraction over δ according to Lemma 1.4 and omit the terms that vanish due to the Bianchi identity:

$$\bar{N}^{\delta}{}_{\beta\gamma} K_{\delta\alpha} = \bar{g}_{ij} \bar{g}^{a_2 f_1} \left(S^i{}_{a_2 b_1 b_2} S^j{}_{c_2 d_1 d_2} + S^i{}_{c_2 b_1 b_2} S^j{}_{d_1 a_2 d_2} \right) S_{e_1 e_2 f_1 f_2}$$
$$x^{b_1} x^{b_2} x^{d_1} x^{e_1} x^{e_2} \nabla_{\beta} x^{c_2} \nabla_{\gamma} x^{d_2} \nabla_{\alpha} x^{f_2}.$$

The integrability condition (0.3b) is equivalent to the vanishing of the antisymmetrisation of the above tensor in α, β, γ. As before, this can be written as

$$\bar{g}_{ij} \bar{g}_{kl} \left(S^{ik}{}_{b_1 b_2} S^j{}_{c_2 d_1 d_2} + S^i{}_{c_2 b_1 b_2} S^j{}_{d_1}{}^k{}_{d_2} \right) S^l{}_{f_2 e_1 e_2}$$
$$x^{b_1} x^{b_2} x^{d_1} x^{e_1} x^{e_2} u^{[c_2} v^{d_2} w^{f_2]} = 0 \qquad (1.26)$$

$$\forall x \in M, \quad \forall u, v, w \in T_x M.$$

Applying (1.11) to the tensor $x^{b_1} x^{b_2} x^{d_1} x^{e_1} x^{e_2} u^{[c_2} v^{d_2} w^{f_2]}$ and decomposing the hook symmetrisers as in (1.8) yields

$$x^{b_1} x^{b_2} x^{d_1} x^{e_1} x^{e_2} u^{[c_2} v^{d_2} w^{f_2]}$$

$$= \text{constant} \cdot \begin{array}{|c|c|c|c|c|} \hline b_2 & b_1 & d_1 & e_1 & e_2 \\ \hline c_2 \\ \cline{1-1} d_2 \\ \cline{1-1} f_2 \\ \cline{1-1} \end{array} \quad x^{b_1} x^{b_2} x^{d_1} x^{e_1} x^{e_2} u^{[c_2} v^{d_2} w^{f_2]}$$

$$+ \text{constant} \cdot \begin{array}{|c|c|c|c|c|c|} \hline c_2 & b_2 & b_1 & d_1 & e_1 & e_2 \\ \hline d_2 \\ \cline{1-1} f_2 \\ \cline{1-1} \end{array}^{\star} \quad x^{b_1} x^{b_2} x^{d_1} x^{e_1} x^{e_2} u^{[c_2} v^{d_2} w^{f_2]}.$$

The following lemma shows that, when substituted into (1.26), only the first term is relevant.

Lemma 1.10.

$$\begin{array}{|c|c|c|c|c|c|} \hline c_2 & b_2 & b_1 & d_1 & e_1 & e_2 \\ \hline d_2 \\ \cline{1-1} f_2 \\ \cline{1-1} \end{array} \quad \bar{g}_{ij} \bar{g}_{kl}\, S^{ik}{}_{b_1 b_2} S^j{}_{c_2 d_1 d_2} S^l{}_{f_2 e_1 e_2} = 0 \qquad (1.27a)$$

$$\begin{array}{|c|c|c|c|c|c|} \hline c_2 & b_2 & b_1 & d_1 & e_1 & e_2 \\ \hline d_2 \\ \cline{1-1} f_2 \\ \cline{1-1} \end{array} \quad \bar{g}_{ij} \bar{g}_{kl} S^i{}_{c_2 b_1 b_2} S^j{}_{d_1}{}^k{}_{d_2}\, S^l{}_{f_2 e_1 e_2} = 0 \qquad (1.27b)$$

Proof. Expanding the antisymmetriser of the Young symmetriser on the left hand side of (1.27a) yields

$$\boxed{c_2}\boxed{b_2}\boxed{b_1}\boxed{d_1}\boxed{e_1}\boxed{e_2}\, \bar{g}_{ij}\bar{g}_{kl}S^{ik}{}_{b_1b_2} \Big(\; S^{j}{}_{c_2d_1d_2}\, S^{l}{}_{f_2e_1e_2} - S^{j}{}_{c_2d_1f_2}\, S^{l}{}_{d_2e_1e_2}$$

$$+ S^{j}{}_{f_2d_1c_2}\, S^{l}{}_{d_2e_1e_2} - S^{j}{}_{d_2d_1c_2}\, S^{l}{}_{f_2e_1e_2}$$

$$+ S^{j}{}_{d_2d_1f_2}\, S^{l}{}_{c_2e_1e_2} - S^{j}{}_{f_2d_1d_2}\, S^{l}{}_{c_2e_1e_2}\Big).$$

Now regard the parenthesis under complete symmetrisation in b_1, b_2, c_2, d_1, e_1, e_2. The last two terms vanish by the Bianchi identity. Renaming i, j, k, l as k, l, i, j in the third term shows that it cancels the fourth due to the contraction with $\bar{g}_{ij}\bar{g}_{kl}S^{ik}{}_{b_1b_2}$. That the first two also cancel each other can be seen after applying twice the symmetrised Bianchi identity (1.18), once to $S^{j}{}_{c_2d_1d_2}$ and once to $S^{l}{}_{f_2e_1e_2}$.

In the same way, the left hand side of (1.27b), written without terms vanishing by the Bianchi identity, is

$$\boxed{c_2}\boxed{b_2}\boxed{b_1}\boxed{d_1}\boxed{e_1}\boxed{e_2}\, \bar{g}_{ij}\bar{g}_{kl}S^{j}{}_{d_1}{}^{k}{}_{c_2} \Big(S^{i}{}_{d_2b_1b_2}\, S^{l}{}_{f_2e_1e_2} - S^{i}{}_{f_2b_1b_2}\, S^{l}{}_{d_2e_1e_2}\Big).$$

Renaming i, j, k, l as l, k, j, i in the first term shows that this is zero, too. □

Remark 1.11. *Again, the lemma could also be deduced from the symmetry classification of Riemann tensor polynomials [FKWC92].*

We have shown the equivalence of the second integrability condition to

$$\bar{g}_{ij}\bar{g}_{kl}\Big(S^{ik}{}_{b_1b_2}\, S^{j}{}_{c_2d_1d_2} + S^{i}{}_{c_2b_1b_2}\, S^{j}{}_{d_1}{}^{k}{}_{d_2}\Big)S^{l}{}_{f_2e_1e_2}$$

$$\begin{array}{|c|c|c|c|c|}\hline b_2 & b_1 & d_1 & e_1 & e_2 \\\hline\end{array}$$
$$\boxed{c_2}$$
$$\boxed{d_2}\qquad x^{b_1}x^{b_2}x^{d_1}x^{e_1}x^{e_2}u^{c_2}v^{d_2}w^{f_2} = 0$$
$$\boxed{f_2}$$

$$\forall x \in M, \quad \forall u, v, w \in T_xM.$$

As before, the restrictions on the vectors u, v, w and x can be dropped, which allows us to write this condition independently of $x, u, v, w \in V$ as

$$
\boxed{\begin{array}{|c|c|c|c|c|}\hline b_2 & b_1 & d_1 & e_1 & e_2 \\\hline c_2 & & & & \\\hline d_2 & & & & \\\hline f_2 & & & & \\\hline\end{array}}^{\star}
\bar{g}_{ij}\bar{g}_{kl}\left(S^{ik}{}_{b_1 b_2}S^{j}{}_{c_2 d_1 d_2} + S^{i}{}_{c_2 b_1 b_2}S^{j}{}_{d_1}{}^{k}{}_{d_2}\right)S^{l}{}_{f_2 e_1 e_2} = 0.
$$

$$(1.28)$$

In order to simplify this condition we need the following two lemmas. For a better readability we will again underline indices which are antisymmetrised.

Lemma 1.12. *The first integrability condition is equivalent to*

$$
\boxed{\begin{array}{|c|}\hline b_2 \\\hline c_2 \\\hline d_2 \\\hline\end{array}}\boxed{\begin{array}{|c|c|}\hline b_1 & d_1 \\\hline\end{array}}\bar{g}_{ij}\left(S^{i}{}_{b_1}{}^{k}{}_{\underline{b}_2} + 2S^{i}{}_{\underline{b}_2}{}^{k}{}_{b_1}\right)S^{j}{}_{\underline{c}_2 d_1 \underline{d}_2} = 0.
$$

$$(1.29)$$

Proof. Take the first integrability condition in the form (1.21c) and reduce the antisymmetriser by the label a_2:

$$
\boxed{\begin{array}{|c|}\hline b_2 \\\hline c_2 \\\hline d_2 \\\hline\end{array}}\bar{g}_{ij}\big(S^{i}{}_{a_2 b_1 \underline{b}_2}S^{j}{}_{\underline{c}_2 d_1 \underline{d}_2} - S^{i}{}_{\underline{b}_2 b_1 \underline{c}_2}S^{j}{}_{\underline{d}_2 d_1 a_2}
$$

$$
+ S^{i}{}_{\underline{c}_2 b_1 \underline{d}_2}S^{j}{}_{a_2 d_1 \underline{b}_2} - S^{i}{}_{\underline{d}_2 b_1 a_2}S^{j}{}_{\underline{b}_2 d_1 \underline{c}_2}\big) = 0.
$$

If we symmetrise this expression in b_1, d_1, the first and third as well as the second and fourth term become equal. Permuting indices, we get

$$
\boxed{\begin{array}{|c|}\hline b_2 \\\hline c_2 \\\hline d_2 \\\hline\end{array}}\boxed{\begin{array}{|c|c|}\hline b_1 & d_1 \\\hline\end{array}}\bar{g}_{ij}\big(S^{i}{}_{a_2 b_1 \underline{b}_2} - S^{i}{}_{\underline{b}_2 a_2 b_1}\big)S^{j}{}_{\underline{c}_2 d_1 \underline{d}_2} = 0.
$$

$$(1.30)$$

If we now symmetrise in a_2, b_1, d_1 and apply the symmetrised Bianchi identity to $S^{i}{}_{a_2 b_1 \underline{b}_2}$, we get back the first integrability condition in the form (1.21b). This proves its equivalence to (1.30). Applying now the Bianchi identity to the first term in (1.30) yields (1.29) with the index k lowered and renamed as a_2. $\qquad\square$

Lemma 1.13. *The following identity is a consequence of the first integrability condition.*

$$\young(c_2,d_2)\;\young(b_2,b_1,d_1)\;\bar{g}_{ij}\left(S^i{}_{b_1}{}^k{}_{b_2}S^j{}_{\underline{c}_2d_1\underline{d}_2} - \tfrac{1}{2}S^i{}_{d_1}{}^k{}_{\underline{d}_2}S^j{}_{\underline{c}_2b_1b_2} - S^i{}_{\underline{d}_2}{}^k{}_{d_1}S^j{}_{\underline{c}_2b_1b_2}\right) = 0$$

$$(1.31)$$

Proof. Reduce the antisymmetriser in (1.29) by the index b_2,

$$\young(c_2,d_2)\;\young(b_1,d_1)\;\bar{g}_{ij}\Big(\;S^i{}_{b_1}{}^k{}_{b_2}S^j{}_{\underline{c}_2d_1\underline{d}_2} + 2S^i{}_{b_2}{}^k{}_{b_1}S^j{}_{\underline{c}_2d_1\underline{d}_2}$$
$$+ S^i{}_{b_1}{}^k{}_{\underline{c}_2}S^j{}_{\underline{d}_2d_1b_2} + 2S^i{}_{\underline{c}_2}{}^k{}_{b_1}S^j{}_{\underline{d}_2d_1b_2}$$
$$+ S^i{}_{b_1}{}^k{}_{\underline{d}_2}S^j{}_{b_2d_1\underline{c}_2} + 2S^i{}_{\underline{d}_2}{}^k{}_{b_1}S^j{}_{b_2d_1\underline{c}_2}\Big) = 0,$$

and then symmetrise in b_2, b_1, d_1. In the last line we can then apply the symmetrised Bianchi identity in order to move the antisymmetrised index c_2 from the fourth to the second position:

$$\young(c_2,d_2)\;\young(b_2,b_1,d_1)\;\bar{g}_{ij}\Big(\;\;S^i{}_{b_1}{}^k{}_{b_2}S^j{}_{\underline{c}_2d_1\underline{d}_2} + 2S^i{}_{b_2}{}^k{}_{b_1}S^j{}_{\underline{c}_2d_1\underline{d}_2}$$
$$- S^i{}_{b_1}{}^k{}_{\underline{d}_2}S^j{}_{\underline{c}_2d_1b_2} - 2S^i{}_{\underline{d}_2}{}^k{}_{b_1}S^j{}_{\underline{c}_2d_1b_2}$$
$$- \tfrac{1}{2}S^i{}_{b_1}{}^k{}_{\underline{d}_2}S^j{}_{\underline{c}_2d_1b_2} - S^i{}_{\underline{d}_2}{}^k{}_{b_1}S^j{}_{\underline{c}_2d_1b_2}\Big) = 0.$$

After permuting indices under symmetrisation appropriately, we get the desired result. $\square$

Proposition 1.14. *Suppose a Killing tensor on a constant curvature manifold satisfies the first Nijenhuis integrability condition* (0.3a). *Then any of the following conditions is equivalent to the second Nijenhuis integrability condition* (0.3b).

(i) The corresponding symmetrised algebraic curvature tensor S satisfies one of the following two equivalent conditions.

$$\young(b_2,c_2,d_2,f_2)\young(b_1,d_1,e_1,e_2)^{\star}\;\;\bar{g}_{ij}\bar{g}_{kl}S^i{}_{c_2d_1d_2}S^j{}_{b_1}{}^k{}_{b_2}S^l{}_{f_2e_1e_2} = 0 \qquad (1.32\text{a})$$

$$\young(b_2,c_2,d_2,f_2)\young(b_1,d_1,e_1,e_2)^{\star}\;\;\bar{g}_{ij}\bar{g}_{kl}\,S^i{}_{c_2b_1b_2}S^j{}_{d_1}{}^k{}_{d_2}S^l{}_{f_2e_1e_2} = 0 \qquad (1.32\text{b})$$

(ii) The corresponding symmetrised algebraic curvature tensor S satisfies one of the following two equivalent conditions.

$$\young(c_2,d_2,f_2)\,\young(b_2b_1d_1e_1e_2)\;\bar g_{ij}\bar g_{kl}S^i{}_{\underline{c}_2 d_1\underline{d}_2}S^j{}_{b_1}{}^k{}_{\underline{b}_2}S^l{}_{\underline{f}_2 e_1 e_2}=0 \qquad (1.33a)$$

$$\young(c_2,d_2,f_2)\,\young(b_2b_1d_1e_1e_2)\;\bar g_{ij}\bar g_{kl}S^i{}_{\underline{c}_2 b_1 b_2}S^j{}_{d_1}{}^k{}_{\underline{d}_2}S^l{}_{\underline{f}_2 e_1 e_2}=0 \qquad (1.33b)$$

(iii) The corresponding symmetrised algebraic curvature tensor S satisfies one of the following three equivalent conditions.

$$\young(b_2,c_2,d_2,f_2)\,\young(b_1d_1e_1e_2)\;\begin{cases}\bar g_{ij}\bar g_{kl}S^i{}_{\underline{c}_2 d_1\underline{d}_2}S^j{}_{b_1}{}^k{}_{\underline{b}_2}S^l{}_{\underline{f}_2 e_1 e_2}=0 & (1.34a)\\[2mm] \bar g_{ij}\bar g_{kl}S^i{}_{\underline{c}_2 d_1\underline{d}_2}S^j{}_{\underline{b}_2}{}^k{}_{b_1}S^l{}_{\underline{f}_2 e_1 e_2}=0 & (1.34b)\\[2mm] \bar g_{ij}\bar g_{kl}S^i{}_{\underline{c}_2 d_1\underline{d}_2}S^j{}_{e_1}{}^k{}_{e_2}S^l{}_{\underline{f}_2 b_1\underline{b}_2}=0 & (1.34c)\end{cases}$$

(iv) The corresponding algebraic curvature tensor R satisfies

$$\young(b_2,c_2,d_2,f_2)\,\young(b_1d_1e_1e_2)\;\bar g_{ij}\bar g_{kl}R^i{}_{d_1\underline{c}_2\underline{d}_2}R^j{}_{e_1}{}^k{}_{e_2}R^l{}_{b_1\underline{f}_2\underline{b}_2}=0. \qquad (1.35)$$

Proof. (i) Contract (1.31) with $\bar g_{kl}S^l{}_{f_2 e_1 e_2}$, antisymmetrise in c_2, d_2, f_2 and symmetrise in b_2, b_1, d_1, e_1, e_2. This yields

$$\young(c_2,d_2,f_2)\,\young(b_2b_1d_1e_1e_2)\;\bar g_{ij}\bar g_{kl}\Big(S^i{}_{b_1}{}^k{}_{b_2}S^j{}_{\underline{c}_2 d_1\underline{d}_2}-\tfrac12 S^i{}_{d_1}{}^k{}_{\underline{d}_2}S^j{}_{\underline{c}_2 b_1 b_2}$$
$$-S^i{}_{\underline{d}_2}{}^k{}_{d_1}S^j{}_{\underline{c}_2 b_1 b_2}\Big)S^l{}_{\underline{f}_2 e_1 e_2}=0.$$

Exchanging d_1 and d_2 in the third term is tantamount to exchanging the upper indices i and k, due to the pair symmetry of $S^i{}_{d_2}{}^k{}_{d_1}$. But under contraction with $\bar g_{ij}\bar g_{kl}$ this is tantamount to exchanging the upper indices j and l. This in turn is tantamount to exchanging c_2, b_1, b_2 with f_2, e_1, e_2 which, under symmetrisation and antisymmetrisation, is tantamount to a sign change. Therefore

$$\young(c_2,d_2,f_2)\,\young(b_2b_1d_1e_1e_2)\;\bar g_{ij}\bar g_{kl}\Big(S^i{}_{b_1}{}^k{}_{b_2}S^j{}_{\underline{c}_2 d_1\underline{d}_2}+\tfrac12 S^i{}_{d_1}{}^k{}_{\underline{d}_2}S^j{}_{\underline{c}_2 b_1 b_2}\Big)S^l{}_{\underline{f}_2 e_1 e_2}=0.$$

$$(1.36)$$

Applying the symmetrised Bianchi identity to $S^i{}_{b_1}{}^k{}_{b_2}$ and antisymmetrising in b_2, c_2, d_2, f_2 yields

$$
\boxed{\begin{array}{|c|c|c|c|}\hline b_2 & b_1 & d_1 & e_1 & e_2 \\\hline c_2 \\\hline d_2 \\\hline f_2 \\\hline\end{array}}^{\star}\; \bar{g}_{ij}\bar{g}_{kl}\left(S^{ik}{}_{b_1 b_2}S^j{}_{c_2 d_1 d_2} - S^i{}_{d_1}{}^k{}_{d_2}S^j{}_{c_2 b_1 b_2}\right)S^l{}_{f_2 e_1 e_2} = 0.
$$

We have derived this identity from the first integrability condition via Lemma 1.13. Comparing it with condition (1.28) shows that (1.28) is equivalent to (1.32b) and, after using once again the symmetrised Bianchi identity, also to (1.32a). This proves (i), since we have already shown that the second integrability condition is equivalent to (1.28).

(ii) Condition (1.33a) is equivalent to (1.32a). This results from (1.11) when taking (1.27a) into account, since the kernels of $PP^\star$ and $P^\star$ coincide:

$$
PP^\star v = 0 \quad \Leftrightarrow \quad \|P^\star v\|^2 = \langle v|PP^\star v\rangle = 0 \quad \Leftrightarrow \quad P^\star v = 0.
$$

In the same way (1.33b) is equivalent to (1.32b), using (1.27b).

(iii) We will prove the equivalence of (1.32a) to each of the equations (1.34). To this aim we establish three linearly independent homogeneous equations for the three tensors on the left hand side of (1.34). For the first equation, we decompose the hook symmetriser in (1.32a) into an antisymmetriser and a symmetriser and then replace the symmetriser by a cyclic sum followed by a symmetrisation in b_1, d_1, e_1, e_2:

$$
\begin{array}{|c|}\hline b_2 \\\hline c_2 \\\hline d_2 \\\hline f_2 \\\hline\end{array}\; \begin{array}{|c|c|c|c|}\hline b_1 & d_1 & e_1 & e_2 \\\hline\end{array}\; \bar{g}_{ij}\bar{g}_{kl}\Big(S^i{}_{\underline{c_2}d_1\underline{d_2}}S^j{}_{b_1}{}^k{}_{\underline{b_2}}S^l{}_{\underline{f_2}e_1 e_2}
$$

$$
+ S^i{}_{\underline{c_2}e_1\underline{d_2}}S^j{}_{d_1}{}^k{}_{b_1}S^l{}_{\underline{f_2}e_2\underline{b_2}}
$$

$$
+ S^i{}_{\underline{c_2}e_2\underline{d_2}}S^j{}_{e_1}{}^k{}_{d_1}S^l{}_{\underline{f_2}\underline{b_2}b_1}
$$

$$
+ S^i{}_{\underline{c_2}\underline{b_2}\underline{d_2}}S^j{}_{e_2}{}^k{}_{e_1}S^l{}_{\underline{f_2}b_1 d_1}
$$

$$
+ S^i{}_{\underline{c_2}b_1\underline{d_2}}S^j{}_{\underline{b_2}}{}^k{}_{e_2}S^l{}_{\underline{f_2}d_1 e_1}\Big) = 0.
$$

The fourth term vanishes by the Bianchi identity and the second term is equal to the third. Therefore (1.32a) is equivalent to

$$
\boxed{\begin{array}{c} b_2 \\ c_2 \\ d_2 \\ f_2 \end{array}} \; \boxed{b_1|d_1|e_1|e_2} \; \bar{g}_{ij}\bar{g}_{kl} S^i{}_{\underline{c}_2 d_1 \underline{d}_2}
$$
$$
\left(S^j{}_{b_1}{}^k{}_{\underline{b}_2} S^l{}_{\underline{f}_2 e_1 e_2} + 2 S^j{}_{e_1}{}^k{}_{e_2} S^l{}_{\underline{f}_2 b_1 \underline{b}_2} + S^j{}_{\underline{b}_2}{}^k{}_{b_1} S^l{}_{\underline{f}_2 e_1 e_2} \right) = 0.
$$
$$(1.37a)$$

This is our first equation. The other two equations follow from the first integrability condition as follows. The second equation is obtained from (1.29) by contracting with $\bar{g}_{kl} S^l{}_{f_2 e_1 e_2}$, antisymmetrising in b_2, c_2, d_2, f_2 and symmetrising in b_1, d_1, e_1, e_2:

$$
\boxed{\begin{array}{c} b_2 \\ c_2 \\ d_2 \\ f_2 \end{array}} \; \boxed{b_1|d_1|e_1|e_2} \; \bar{g}_{ij}\bar{g}_{kl} \left(S^i{}_{b_1}{}^k{}_{\underline{b}_2} + 2 S^i{}_{\underline{b}_2}{}^k{}_{b_1} \right) S^j{}_{\underline{c}_2 d_1 \underline{d}_2} S^l{}_{\underline{f}_2 e_1 e_2} = 0.
$$

This can be rewritten as

$$
\boxed{\begin{array}{c} b_2 \\ c_2 \\ d_2 \\ f_2 \end{array}} \; \boxed{b_1|d_1|e_1|e_2} \; \bar{g}_{ij}\bar{g}_{kl} S^i{}_{\underline{c}_2 d_1 \underline{d}_2} \left(S^j{}_{b_1}{}^k{}_{\underline{b}_2} S^l{}_{\underline{f}_2 e_1 e_2} + 2 S^j{}_{\underline{b}_2}{}^k{}_{b_1} S^l{}_{\underline{f}_2 e_1 e_2} \right) = 0
$$
$$(1.37b)$$

and is our second equation. For the third equation, we rename b_1, b_2 in (1.31) as e_1, e_2, contract with $\bar{g}_{kl} S^l{}_{f_2 b_1 b_2}$, antisymmetrise in b_2, c_2, d_2, f_2 and symmetrise in b_1, d_1, e_1, e_2:

$$
\boxed{\begin{array}{c} b_2 \\ c_2 \\ d_2 \\ f_2 \end{array}} \; \boxed{b_1|d_1|e_1|e_2} \; \bar{g}_{ij}\bar{g}_{kl} S^l{}_{\underline{f}_2 b_1 \underline{b}_2}
$$
$$
\left(S^i{}_{e_1}{}^k{}_{e_2} S^j{}_{\underline{c}_2 d_1 \underline{d}_2} - \tfrac{1}{2} S^i{}_{d_1}{}^k{}_{\underline{d}_2} S^j{}_{\underline{c}_2 e_1 e_2} - S^i{}_{\underline{d}_2}{}^k{}_{d_1} S^j{}_{\underline{c}_2 e_1 e_2} \right) = 0.
$$

This can be rewritten as

$$
\boxed{\begin{array}{c} b_2 \\ c_2 \\ d_2 \\ f_2 \end{array}} \; \boxed{b_1|d_1|e_1|e_2} \; \bar{g}_{ij}\bar{g}_{kl} S^i{}_{\underline{c}_2 d_1 \underline{d}_2}
$$
$$
\left(S^j{}_{e_1}{}^k{}_{e_2} S^l{}_{\underline{f}_2 b_1 \underline{b}_2} - \tfrac{1}{2} S^j{}_{\underline{b}_2}{}^k{}_{b_1} S^l{}_{\underline{f}_2 e_1 e_2} - S^j{}_{b_1}{}^k{}_{\underline{b}_2} S^l{}_{\underline{f}_2 e_1 e_2} \right) = 0
$$
$$(1.37c)$$

and is our last equation. Clearly, the resulting homogeneous system (1.37) implies (1.34). On the other hand, any of the equations (1.34) together with (1.37b) and (1.37c) implies (1.37a) and therefore (1.32a).

(iv) Condition (1.35) is equivalent to (1.34c) via (1.2).   □

1.4 Redundancy of the 3rd integrability condition

With Propositions 1.9 and 1.14 we have proven the equivalence of the first two of the Nijenhuis integrability conditions (0.3) for a Killing tensor on a constant curvature manifold with the two algebraic integrability conditions (0.13) for the associated algebraic curvature tensor. We now finish the proof of Theorem I with the following result. Notice that it is not limited to constant curvature.

Proposition 1.15. *For a Killing tensor on an arbitrary Riemannian manifold the third of the three Nijenhuis integrability conditions* (0.3) *is redundant.*

Proof. For compactness of notation, let us denote just in this proof the covariant derivative of $K_{\alpha\beta}$ by $K_{\alpha\beta,\gamma}$ instead of $\nabla_\gamma K_{\alpha\beta}$.

A Killing tensor is symmetric by definition. Hence at any point of the manifold we can choose an orthonormal basis of the tangent space in which the Killing tensor is diagonal, i.e. $K^\alpha{}_\beta = \lambda_\alpha \delta^\alpha{}_\beta$ (no sum). In this basis, the Nijenhuis torsion (0.2) reads

$$N_{\alpha\beta\gamma} = (\lambda_\alpha - \lambda_\beta)K_{\alpha[\beta,\gamma]}.$$

Substituted into the Nijenhuis integrability conditions (0.3) we get

$$0 = N^\delta{}_{[\beta\gamma}g_{\alpha]\delta} = \boxed{\begin{smallmatrix}\alpha\\\beta\\\gamma\end{smallmatrix}}\,N_{\alpha\beta\gamma} = \boxed{\begin{smallmatrix}\alpha\\\beta\\\gamma\end{smallmatrix}}\,(\lambda_\alpha - \lambda_\beta)K_{\alpha\beta,\gamma}$$

$$0 = N^\delta{}_{[\beta\gamma}K_{\alpha]\delta} = \boxed{\begin{smallmatrix}\alpha\\\beta\\\gamma\end{smallmatrix}}\,\lambda_\alpha N_{\alpha\beta\gamma} = \boxed{\begin{smallmatrix}\alpha\\\beta\\\gamma\end{smallmatrix}}\,\lambda_\alpha(\lambda_\alpha - \lambda_\beta)K_{\alpha\beta,\gamma}$$

$$0 = N^\delta{}_{[\beta\gamma}K_{\alpha]\varepsilon}K^\varepsilon{}_\delta = \boxed{\begin{smallmatrix}\alpha\\\beta\\\gamma\end{smallmatrix}}\,\lambda_\alpha^2 N_{\alpha\beta\gamma} = \boxed{\begin{smallmatrix}\alpha\\\beta\\\gamma\end{smallmatrix}}\,\lambda_\alpha^2(\lambda_\alpha - \lambda_\beta)K_{\alpha\beta,\gamma}.$$

As before, the Young operators stand for a complete antisymmetrisation in the indices α, β and γ. Using that $K_{\alpha\beta,\gamma}$ is symmetric in α, β and that a complete antisymmetrisation over α, β, γ can be split into an antisymmetrisation in α, β and a subsequent sum over the cyclic permutations of α, β, γ, we can rewrite the preceding equations as

$$0 = (\lambda_\alpha - \lambda_\beta)K_{\alpha\beta,\gamma} + \text{cyclic} \tag{1.38a}$$

$$0 = (\lambda_\alpha + \lambda_\beta)(\lambda_\alpha - \lambda_\beta)K_{\alpha\beta,\gamma} + \text{cyclic} \tag{1.38b}$$

$$0 = (\lambda_\alpha^2 + \lambda_\beta^2)(\lambda_\alpha - \lambda_\beta)K_{\alpha\beta,\gamma} + \text{cyclic}, \tag{1.38c}$$

where "+cyclic" stands for a summation over the cyclic permutations of α, β, γ. These equations are one by one equivalent to the Nijenhuis integrability conditions (0.3). In the same way we can write the Killing equation as

$$0 = K_{\alpha\beta,\gamma} + \text{cyclic}. \tag{1.39}$$

We want to prove that (1.39) together with $(1.38a)$ and $(1.38b)$ imply $(1.38c)$. To this end we write the first three equations for $K_{\alpha\beta,\gamma}$ in matrix form as

$$\begin{bmatrix} 1 & 1 & 1 \\ \lambda_\alpha - \lambda_\beta & \lambda_\beta - \lambda_\gamma & \lambda_\gamma - \lambda_\alpha \\ \lambda_\alpha^2 - \lambda_\beta^2 & \lambda_\beta^2 - \lambda_\gamma^2 & \lambda_\gamma^2 - \lambda_\alpha^2 \end{bmatrix} \begin{bmatrix} K_{\alpha\beta,\gamma} \\ K_{\beta\gamma,\alpha} \\ K_{\gamma\alpha,\beta} \end{bmatrix} = 0.$$

The determinant of the coefficient matrix is a multiple of the Vandermode determinant. If the eigenvalues $\lambda_\alpha, \lambda_\beta, \lambda_\gamma$ are pairwise different, this implies that $K_{\alpha\beta,\gamma} = K_{\beta\gamma,\alpha} = K_{\gamma\alpha,\beta} = 0$. If exactly two of the eigenvalues are equal, say $\lambda_\alpha \neq \lambda_\beta = \lambda_\gamma$, then we have $K_{\alpha\beta,\gamma} = -\frac{1}{2}K_{\beta\gamma,\alpha} = K_{\gamma\alpha,\beta}$. For three equal eigenvalues the only restriction on $K_{\alpha\beta,\gamma}$ is the Killing equation $K_{\alpha\beta,\gamma} + K_{\beta\gamma,\alpha} + K_{\gamma\alpha,\beta} = 0$. In all three cases we see, that the equation $(1.38c)$ is also satisfied. $\quad\square$

1.5 Commuting Killing tensors

In the following we translate the condition that two Killing tensors commute as endomorphisms into a purely algebraic condition on their associated algebraic curvature tensors. Notice that this condition is *not* given, as one may guess naïvely, by the vanishing of the commutator of the algebraic curvature tensors as endomorphisms on 2-forms.

Proposition 1.16. *Let K and $\tilde{K}$ be two Killing tensors on a non-flat constant curvature manifold with algebraic curvature tensors R and $\tilde{R}$. Then the following statements are equivalent.*

$$[K, \tilde{K}] = 0 \tag{1.40a}$$

$$\boxed{\begin{array}{|c|c|c|c|}\hline b_1 & b_2 & d_1 & d_2 \\\hline a_2 & & & \\\hline c_2 & & & \\\hline\end{array}}^{\star} g_{ij} R^i{}_{b_1 a_2 b_2} \tilde{R}^j{}_{d_1 c_2 d_2} = 0 \tag{1.40b}$$

$$\boxed{\begin{array}{|c|}\hline a_2 \\\hline c_2 \\\hline\end{array}}\ \boxed{\begin{array}{|c|c|c|c|}\hline b_1 & b_2 & d_1 & d_2 \\\hline\end{array}}\ g_{ij} R^i{}_{b_1 a_2 b_2} \tilde{R}^j{}_{d_1 c_2 d_2} = 0 \tag{1.40c}$$

Proof. Using $\nabla_v x^b = v^b$, we write (0.7) in local coordinates for coordinate vectors $v = \partial_\alpha$ and $w = \partial_\beta$:

$$K_{\alpha\beta} = R_{a_1 b_1 a_2 b_2} x^{a_1} x^{a_2} \nabla_\alpha x^{b_1} \nabla_\beta x^{b_2}.$$

The product of K and $\tilde{K}$, regarded as endomorphisms, is then given by

$$K^\alpha{}_\gamma \tilde{K}^\gamma{}_\beta = R_{a_1 b_1 a_2 b_2} \tilde{R}_{c_1 d_1 c_2 d_2} x^{a_1} x^{a_2} x^{c_1} x^{c_2} \nabla^\alpha x^{b_1} \nabla_\gamma x^{b_2} \nabla^\gamma x^{d_1} \nabla_\beta x^{d_2}.$$

By Lemma 1.4 we have

$$\nabla_\gamma x^{b_2} \nabla^\gamma x^{d_1} = g^{b_2 d_1} - x^{b_2} x^{d_1}.$$

As a consequence of the antisymmetry of algebraic curvature tensors in the last index pair, the term $x^{b_2} x^{d_1}$ does not contribute when substituting this identity into the previous expression:

$$K^\alpha{}_\gamma \tilde{K}^\gamma{}_\beta = g^{b_2 d_1} R_{a_1 b_1 a_2 b_2} \tilde{R}_{c_1 d_1 c_2 d_2} x^{a_1} x^{a_2} x^{c_1} x^{c_2} \nabla^\alpha x^{b_1} \nabla_\beta x^{d_2}.$$

The commutator $[K, \tilde{K}] = K\tilde{K} - \tilde{K}K$ is therefore given by

$$[K, \tilde{K}]_{\alpha\beta} = g^{b_2 d_1} R_{a_1 b_1 a_2 b_2} \tilde{R}_{c_1 d_1 c_2 d_2} x^{a_1} x^{a_2} x^{c_1} x^{c_2} \nabla_{[\alpha} x^{b_1} \nabla_{\beta]} x^{d_2}$$

and vanishes if and only if

$$g^{b_2 d_1} R_{a_1 b_1 a_2 b_2} \tilde{R}_{c_1 d_1 c_2 d_2} x^{a_1} x^{a_2} x^{c_1} x^{c_2} v^{[b_1} w^{d_2]} = 0 \qquad (1.41)$$

for all $x \in M$ and $v, w \in T_x M$. That is, for all $x, v, w \in V$ with

$$g(x, x) = 1 \qquad\qquad g(x, v) = g(x, w) = 0. \qquad (1.42)$$

We can drop the restrictions $g(x, v) = 0$ and $g(x, w) = 0$ by decomposing arbitrary vectors $v, w \in V$ under the decomposition $V = T_x M \oplus \mathbb{R} x$. To see this, notice that (1.41) is trivially satisfied for $v = x$ or for $w = x$. Indeed, in this case the tensor

$$R_{a_1 b_1 a_2 b_2} \tilde{R}_{c_1 d_1 c_2 d_2} \qquad (1.43)$$

is implicitly symmetrised over five indices and Dirichlet's drawer principle tells us that this comprises a symmetrisation in one of the four antisymmetric index pairs. We can also drop the restriction $g(x, x) = 1$, since (1.41) is a homogeneous polynomial in x for fixed v, w and $\mathbb{R} M \subseteq V$ is open. This means that we can omit the restrictions (1.42) completely. In other words, $[K, \tilde{K}] = 0$ is equivalent to (1.41) being satisfied for *all* $x, v, w \in V$.

Applying (1.12) to the tensor $x^{a_1} x^{a_2} x^{c_1} x^{c_2} v^{[b_1} w^{d_2]}$ and decomposing the hook symmetrisers as in (1.8) we get

$$x^{a_1} x^{a_2} x^{c_1} x^{c_2} v^{[b_1} w^{d_2]}$$

$$= \frac{1}{2!} \boxed{\begin{smallmatrix} b_1 \\ d_2 \end{smallmatrix}} \cdot \frac{1}{4!} \boxed{\begin{smallmatrix} a_1 & a_2 & c_1 & c_2 \end{smallmatrix}} x^{a_1} x^{a_2} x^{c_1} x^{c_2} v^{[b_1} w^{d_2]}$$

$$= \left(\frac{4! \cdot 3!}{10368} \boxed{\begin{smallmatrix} a_1 & a_2 & c_1 & c_2 \\ b_1 \\ d_2 \end{smallmatrix}} + \frac{2! \cdot 5!}{34560} \boxed{\begin{smallmatrix} b_1 & a_1 & a_2 & c_1 & c_2 \\ d_2 \end{smallmatrix}}^{\star} \right) x^{a_1} x^{a_2} x^{c_1} x^{c_2} v^{b_1} w^{d_2}.$$

When substituted into (1.41), this yields

$$g^{b_2 d_1} R_{a_1 b_1 a_2 b_2} \tilde{R}_{c_1 d_1 c_2 d_2}$$

$$\left(\left(\frac{1}{72} \boxed{\begin{smallmatrix} a_1 & a_2 & c_1 & c_2 \\ b_1 \\ d_2 \end{smallmatrix}} + \frac{1}{144} \boxed{\begin{smallmatrix} b_1 & a_1 & a_2 & c_1 & c_2 \\ d_2 \end{smallmatrix}}^{\star} \right) x^{a_1} x^{a_2} x^{c_1} x^{c_2} v^{b_1} w^{d_2} \right) = 0.$$

As before, we can replace the Young symmetrisers acting on upper indices by their corresponding adjoints acting on the corresponding lower indices:

$$\left(\left(\frac{1}{72}\;\boxed{\begin{array}{cccc}a_1&a_2&c_1&c_2\\\hline b_1\\\cline{1-1}d_2\end{array}}^{\;\star} + \frac{1}{144}\;\boxed{\begin{array}{ccccc}b_1&a_1&a_2&c_1&c_2\\\hline d_2\end{array}}\right)g^{b_2 d_1}R_{a_1 b_1 a_2 b_2}\tilde{R}_{c_1 d_1 c_2 d_2}\right)$$
$$x^{a_1}x^{a_2}x^{c_1}x^{c_2}v^{b_1}w^{d_2} = 0.$$

Now notice that the second Young tableau involves a symmetrisation over the five indices b_1, a_1, a_2, c_1, c_2 and that, as above, the symmetrisation of the tensor (1.43) in any five indices is zero. Hence the second term in the last equation vanishes and we obtain

$$\left(\boxed{\begin{array}{cccc}a_1&a_2&c_1&c_2\\\hline b_1\\\cline{1-1}d_2\end{array}}^{\;\star}g^{b_2 d_1}R_{a_1 b_1 a_2 b_2}\tilde{R}_{c_1 d_1 c_2 d_2}\right)x^{a_1}x^{a_2}x^{c_1}x^{c_2}v^{b_1}w^{d_2} = 0.$$

Recall that $[K, \tilde{K}] = 0$ is equivalent to this condition being satisfied for all $x, v, w \in V$. By polarising in x we get

$$\boxed{\begin{array}{cccc}a_1&a_2&c_1&c_2\\\hline b_1\\\cline{1-1}d_2\end{array}}^{\;\star}g^{b_2 d_1}R_{a_1 b_1 a_2 b_2}\tilde{R}_{c_1 d_1 c_2 d_2} = 0.$$

This is the same as (1.40b) after appropriately renaming, lowering and rising indices. We have proven the equivalence (1.40a) $\Leftrightarrow$ (1.40b).

We now prove the equivalence (1.40b) $\Leftrightarrow$ (1.40c). Start from (1.40b) by expanding the Young tableau:

$$\boxed{\begin{array}{c}b_1\\\hline a_2\\\hline c_2\end{array}}\;\boxed{\begin{array}{cccc}b_1&b_2&d_1&d_2\end{array}}g_{ij}R^i{}_{b_1 a_2 b_2}\tilde{R}^j{}_{d_1 c_2 d_2} = 0.$$

In order to sum over all 4! permutations when carrying out the symmetrisation in the indices b_1, b_2, d_1, d_2, one can first take the sum over the 4 cyclic permutations of b_1, b_2, d_1, d_2, then fix the index b_1

and finally sum over all 3! permutations of the remaining 3 indices b_2, d_1, d_2:

$$\boxed{\begin{smallmatrix}b_1\\a_2\\c_2\end{smallmatrix}}\;\boxed{b_2\,|\,d_1\,|\,d_2}\; g_{ij}\left(R^i{}_{\underline{b}_1\underline{a}_2 b_2}\tilde{R}^j{}_{d_1\underline{c}_2 d_2} + R^i{}_{b_2\underline{a}_2 d_1}\tilde{R}^j{}_{d_2\underline{c}_2\underline{b}_1}\right.$$

$$\left. + R^i{}_{d_1\underline{a}_2 d_2}\tilde{R}^j{}_{\underline{b}_1\underline{c}_2 b_2} + R^i{}_{d_2\underline{a}_2\underline{b}_1}\tilde{R}^j{}_{b_2\underline{c}_2 d_1}\right) = 0.$$

As before, we underlined each antisymmetrised index for a better readability. Permuting the indices of the terms in the parenthesis under symmetrisation in b_2, d_1, d_2 and antisymmetrisation in b_1, a_2, c_2 we can gather the first and last as well as the second and third term:

$$\boxed{\begin{smallmatrix}b_1\\a_2\\c_2\end{smallmatrix}}\;\boxed{b_2\,|\,d_1\,|\,d_2}\; g_{ij}\left((R^i{}_{\underline{b}_1\underline{a}_2 b_2} + R^i{}_{b_2\underline{a}_2\underline{b}_1})\tilde{R}^j{}_{d_1\underline{c}_2 d_2} + \right.$$

$$\left. R^i{}_{d_1\underline{a}_2 d_2}(\tilde{R}^j{}_{\underline{b}_1\underline{c}_2 b_2} + \tilde{R}^j{}_{b_2\underline{c}_2\underline{b}_1})\right) = 0.$$

Using the symmetries of R, the terms in the inner parentheses can be rewritten as

$$R^i{}_{\underline{b}_1\underline{a}_2 b_2} = -R^i{}_{\underline{b}_1 b_2\underline{a}_2}$$

$$R^i{}_{b_2\underline{a}_2\underline{b}_1} = -R^i{}_{\underline{a}_2\underline{b}_1 b_2} - R^i{}_{\underline{b}_1 b_2\underline{a}_2} = R^i{}_{\underline{a}_2 b_2\underline{b}_1} - R^i{}_{\underline{b}_1 b_2\underline{a}_2}$$

resulting in

$$\boxed{\begin{smallmatrix}b_1\\a_2\\c_2\end{smallmatrix}}\;\boxed{b_2\,|\,d_1\,|\,d_2}\; g_{ij}\left(R^i{}_{\underline{a}_2 b_2\underline{b}_1}\tilde{R}^j{}_{d_1\underline{c}_2 d_2} + R^i{}_{d_1\underline{a}_2 d_2}\tilde{R}^j{}_{\underline{c}_2 b_2\underline{b}_1}\right) = 0.$$

As above, when carrying out the antisymmetrisation over b_1, a_2, c_2, we can first sum over the three cyclic permutations of b_1, a_2, c_2, then fix b_1 and finally sum over the two permutations of a_2, c_2. This results in

$$\boxed{\begin{smallmatrix}a_2\\c_2\end{smallmatrix}}\;\boxed{b_2\,|\,d_1\,|\,d_2}\; g_{ij}\left(R^i{}_{a_2 b_2 b_1}\tilde{R}^j{}_{d_1 c_2 d_2} + R^i{}_{d_1 a_2 d_2}\tilde{R}^j{}_{c_2 b_2 b_1}\right.$$

$$+ R^i{}_{c_2 b_2 a_2}\tilde{R}^j{}_{d_1 b_1 d_2} + R^i{}_{d_1 c_2 d_2}\tilde{R}^j{}_{b_1 b_2 a_2}$$

$$\left. + R^i{}_{b_1 b_2 c_2}\tilde{R}^j{}_{d_1 a_2 d_2} + R^i{}_{d_1 b_1 d_2}\tilde{R}^j{}_{a_2 b_2 c_2}\right) = 0.$$

The symmetrisation of this in b_1, b_2, d_1, d_2 yields

$$\boxed{\begin{smallmatrix}a_2\\c_2\end{smallmatrix}}\;\boxed{b_1\,|\,b_2\,|\,d_1\,|\,d_2}\;g_{ij}\left(R^i{}_{\underline{a_2}b_2b_1}\tilde{R}^j{}_{d_1\underline{c_2}d_2} + R^i{}_{d_1\underline{a_2}d_2}\tilde{R}^j{}_{\underline{c_2}b_2b_1}\right.$$
$$+ R^i{}_{\underline{c_2}b_2\underline{a_2}}\tilde{R}^j{}_{d_1b_1d_2} + R^i{}_{d_1\underline{c_2}d_2}\tilde{R}^j{}_{b_1b_2\underline{a_2}}$$
$$\left.+ R^i{}_{b_1b_2\underline{c_2}}\tilde{R}^j{}_{d_1\underline{a_2}d_2} + R^i{}_{d_1b_1d_2}\tilde{R}^j{}_{\underline{a_2}b_2\underline{c_2}}\right) = 0,$$

where we again underlined antisymmetrised indices. By the antisymmetry of algebraic curvature tensors in the second index pair, all but the fourth and fifth term in the parenthesis vanish and we get

$$\boxed{\begin{smallmatrix}a_2\\c_2\end{smallmatrix}}\;\boxed{b_1\,|\,b_2\,|\,d_1\,|\,d_2}\;g_{ij}\left(R^i{}_{d_1\underline{c_2}d_2}\tilde{R}^j{}_{b_1\underline{a_2}b_2} + R^i{}_{b_1\underline{c_2}b_2}\tilde{R}^j{}_{d_1\underline{a_2}d_2}\right) = 0.$$

Due to the symmetrisation and antisymmetrisation we can permute the indices of the terms inside the parenthesis to find that

$$\boxed{\begin{smallmatrix}a_2\\c_2\end{smallmatrix}}\;\boxed{b_1\,|\,b_2\,|\,d_1\,|\,d_2}\;g_{ij}R^i{}_{b_1\underline{a_2}b_2}\tilde{R}^j{}_{d_1\underline{c_2}d_2} = 0.$$

Recall that this equation has been obtained from (1.40b) by a symmetrisation. This proves (1.40b) $\Rightarrow$ (1.40c). The converse follows easily by antisymmetrising (1.40c) in b_1, a_2, c_2. This achieves the proof of the theorem. $\square$

Remark 1.17. *Obviously* (1.40b) *must be true for* $\tilde{R} = R$. *This fact is not evident and we leave it to the reader as an exercise for the manipulations used above.*

For diagonal algebraic curvature tensors in dimension four we can rewrite condition (1.40c) as a determinant.

Corollary 1.18. *Two Killing tensors on a 3-dimensional (non-flat) constant curvature manifold with diagonal algebraic curvature tensors R and $\tilde{R}$ commute if and only if*

$$\det\begin{pmatrix}1 & R_{ijij} & \tilde{R}_{ijij}\\ 1 & R_{jkjk} & \tilde{R}_{jkjk}\\ 1 & R_{kiki} & \tilde{R}_{kiki}\end{pmatrix} = 0 \qquad (1.44)$$

for all pairwise distinct $i, j, k \in \{0, 1, 2, 3\}$.

2 The proof of concept: a complete solution for the 3-dimensional sphere

The art of doing mathematics
consists in finding that special case
which contains all the germs of generality.

DAVID HILBERT *(1873 – 1943)*

Contents

This chapter is dedicated to proving Theorem II and its consequences.

2.1 Properties of algebraic curvature tensors

Given a scalar product g on V, we can raise and lower indices. The symmetries (0.6a) and (0.6b) then allow us to regard an algebraic curvature tensor $R_{a_1 a_2 b_1 b_2}$ on V as a symmetric endomorphism $R^{a_1 b_1}{}_{a_2 b_2}$ on the space $\Lambda^2 V$ of 2-forms on V. Since we will frequently change between both interpretations, we denote endomorphisms by the same letter in boldface.

2.1.1 Decomposition

In the special case where $\dim V = 4$, the Hodge star operator "$*$" defines a decomposition

$$\Lambda^2 V = \Lambda_+^2 V \oplus \Lambda_-^2 V$$

of $\Lambda^2 V$ into its ± 1 eigenspaces. We can therefore write an algebraic curvature tensor R and the Hodge star as block matrices

$$\mathbf{R} = \left(\begin{array}{c|c} W_+ & T_- \\ \hline T_+ & W_- \end{array} \right) + \frac{s}{12} \left(\begin{array}{c|c} I_+ & 0 \\ \hline 0 & I_- \end{array} \right) \quad * = \left(\begin{array}{c|c} +I_+ & 0 \\ \hline 0 & -I_- \end{array} \right)$$

$$(2.1a)$$

with the 3×3-blocks satisfying

$$W_+^t = W_+ \qquad W_-^t = W_- \qquad T_- = T_+^t \qquad (2.1b)$$

$$\operatorname{tr} W_+ + \operatorname{tr} W_- = 0 \qquad (2.1c)$$

$$\operatorname{tr} W_+ - \operatorname{tr} W_- = 0. \qquad (2.1d)$$

Here I_+ and I_- denote the identity on $\Lambda^2_+ V$ respectively $\Lambda^2_- V$. The conditions (2.1b) assure symmetry, condition (2.1c) says that $s = 2\operatorname{tr}\mathbf{R}$ and condition (2.1d) is a reformulation of the Bianchi identity (0.6c). Indeed, the Bianchi identity is equivalent to the vanishing of the antisymmetrisation of $R_{a_1 b_1 a_2 b_2}$ in all four indices. In dimension four this can be written as $\varepsilon^{a_2 b_2 a_1 b_1} R_{a_1 b_1 a_2 b_2} = 0$ or

$$\varepsilon^{a_2 b_2}{}_{a_1 b_1} \mathbf{R}^{a_1 b_1}{}_{a_2 b_2} = 0,$$

where $\varepsilon^{a_2 b_2 a_1 b_1}$ is the totally antisymmetric tensor. Remark that $\varepsilon^{a_2 b_2}{}_{a_1 b_1}$ is nothing else than the Hodge star operator and hence

$$\operatorname{tr}(*\mathbf{R}) = 0, \tag{2.2}$$

where $\operatorname{tr} : \operatorname{End}(\Lambda^2 V) \to \mathbb{R}$ is the usual trace. Now (2.1d) is (2.2) applied to $\mathbf{R}$ in (2.1a).

The space of algebraic curvature tensors is an irreducible $\operatorname{GL}(V)$-representation and (2.1) gives a decomposition of this representation into irreducible representations of the subgroup $\operatorname{SO}(V)$ when $\dim V = 4$. As the notation already suggests, we can relate these components to the familiar Ricci decomposition

$$\mathbf{R} = \mathbf{W} + \mathbf{T} + \mathbf{S}$$

of an algebraic curvature tensor R into three parts:

- The *scalar part*, given by the *scalar curvature s*:

$$S_{a_1 b_1 a_2 b_2} = \tfrac{s}{12}(g_{a_1 a_2} g_{b_1 b_2} - g_{a_1 b_2} g_{b_1 a_2}) \quad s = g^{a_1 a_2} g^{b_1 b_2} R_{a_1 b_1 a_2 b_2}$$

- The *trace free Ricci part*, given by the *trace free Ricci tensor* $T_{a_1 a_2}$:

$$T_{a_1 b_1 a_2 b_2} = \tfrac{1}{2}(T_{a_1 a_2} g_{b_1 b_2} - T_{a_1 b_2} g_{b_1 a_2} - T_{b_1 a_2} g_{a_1 b_2} + T_{b_1 b_2} g_{a_1 a_2})$$
$$T_{a_1 a_2} = g^{b_1 b_2} R_{a_1 b_1 a_2 b_2} - \tfrac{s}{4} g_{a_1 a_2}$$

- The *Weyl part*, given by the totally trace free *Weyl tensor* $W := R - T - S$.

It is not difficult to check that $[*, \mathbf{T}] = 0$ and hence

$$\mathbf{W} = \left(\begin{array}{c|c} W_+ & 0 \\ \hline 0 & W_- \end{array} \right) \quad \mathbf{T} = \left(\begin{array}{c|c} 0 & T_- \\ \hline T_+ & 0 \end{array} \right) \quad \mathbf{S} = \frac{s}{12} \left(\begin{array}{c|c} I_+ & 0 \\ \hline 0 & I_- \end{array} \right).$$

(2.3)

We see that W_+ and W_- are the self-dual and anti-self-dual part of the Weyl tensor. Implicitly, the above interpretation also provides an isomorphism

$$\begin{array}{ccc} S_0^2 V & \xrightarrow{\cong} & \operatorname{Hom}(\Lambda_+^2 V, \Lambda_-^2 V) \\ T & \mapsto & T_+ \end{array}$$

(2.4)

between trace-free Ricci tensors and homomorphisms from self-dual to anti-self-dual 2-forms.

Remark 2.1. *The integrability of a Killing tensor does not depend on the scalar part of its associated algebraic curvature tensor, because this part corresponds to a multiple of the metric. This is the reason why throughout this exposition we can safely ignore the scalar part or set it to any convenient value. Sometimes we will also absorb it in the Weyl part by waiving the trace condition* (2.1c).

2.1.2 The action of the isometry group

Recall that the isometry group of $\mathbb{S}^n \subset V$ is $O(V)$ and that $SO(V)$ is the subgroup of orientation preserving isometries. We first examine the induced action of $SO(V)$ on algebraic curvature tensors of the form (2.1). The standard action of $SO(V)$ on V induces a natural action on $\Lambda^2 V$ which is the adjoint action of $SO(V)$ on its Lie algebra under the isomorphism

$$\mathfrak{so}(V) \cong \Lambda^2 V.$$

Now consider the following commutative diagram of Lie group morphisms for $\dim V = 4$.

$$\begin{array}{ccc}
\mathrm{Spin}(V) & \xrightarrow{\;\cong\;} & \mathrm{Spin}(\Lambda^2_+ V) \times \mathrm{Spin}(\Lambda^2_- V) \\
\downarrow{\scriptstyle 2:1} & & \downarrow{\scriptstyle 2:1} \qquad \downarrow{\scriptstyle 2:1} \\
\pi \colon \mathrm{SO}(V) & \xrightarrow{\;2:1\;} & \mathrm{SO}(\Lambda^2_+ V) \times \mathrm{SO}(\Lambda^2_- V) \hookrightarrow \mathrm{SO}(\Lambda^2 V) \\
U & \longmapsto & (U_+, U_-) \qquad\qquad \mapsto \begin{pmatrix} U_+ & 0 \\ 0 & U_- \end{pmatrix}.
\end{array}$$

Here the double covering π in the second row is induced from the exceptional isomorphism in the first row via the universal covering maps (vertical). Under the induced isomorphism of Lie algebras,

$$\pi_* \colon \mathfrak{so}(V) \xrightarrow{\;\cong\;} \mathfrak{so}(\Lambda^2_+ V) \oplus \mathfrak{so}(\Lambda^2_- V),$$

the adjoint action of an element $U \in \mathrm{SO}(V)$ corresponds to the adjoint actions of $U_+ \in \mathrm{SO}(\Lambda^2_+ V)$ and $U_- \in \mathrm{SO}(\Lambda^2_- V)$. Since $\Lambda^2_+ V$ has dimension three, the Hodge star operator gives an isomorphism

$$\mathfrak{so}(\Lambda^2_+ V) \cong \Lambda^2 \Lambda^2_+ V \cong \Lambda^2_+ V$$

under which the adjoint action of U_+ corresponds to the standard action on $\Lambda^2_+ V$ and similarly for U_-. The Hodge decomposition completes the above isomorphisms to a commutative diagram

$$\begin{array}{ccc}
\pi_* \colon \mathfrak{so}(V) & \xrightarrow{\;\cong\;} & \mathfrak{so}(\Lambda^2_+ V) \oplus \mathfrak{so}(\Lambda^2_- V) \\
\downarrow{\scriptstyle \cong} & & \downarrow{\scriptstyle \cong} \qquad \downarrow{\scriptstyle \cong} \\
\Lambda^2 V & \xrightarrow{\;\cong\;} & \Lambda^2_+ V \quad \oplus \quad \Lambda^2_- V,
\end{array}$$

which shows that the natural action of $U \in \mathrm{SO}(V)$ on $\Lambda^2 V$ is given by

$$\begin{pmatrix} U_+ & 0 \\ 0 & U_- \end{pmatrix} \in \pi\big(\mathrm{SO}(V)\big) \subset \mathrm{SO}(\Lambda^2 V). \tag{2.5}$$

Hence the natural action of U on $\mathrm{End}(\Lambda^2 V)$ is given by conjugation with this matrix. Restricting to algebraic curvature tensors we get:

Proposition 2.2. *Let $\pi(U) = (U_+, U_-)$ be the image of an element $U \in \mathrm{SO}(V)$ under the double covering*

$$\pi \colon \mathrm{SO}(V) \to \mathrm{SO}(\Lambda_+^2 V) \times \mathrm{SO}(\Lambda_-^2 V).$$

Then U acts on algebraic curvature tensors in the form (2.1) by conjugation with (2.5), i.e. via

$$W_+ \mapsto U_+^t W_+ U_+ \qquad W_- \mapsto U_-^t W_- U_- \qquad T_+ \mapsto U_-^t T_+ U_+ \qquad (2.6)$$

and trivially on the scalar part.

We can describe the twofold cover π explicitly in terms of orthonormal bases on the spaces V, $\Lambda_+^2 V$ and $\Lambda_-^2 V$. It maps an orthonormal basis (e_0, e_1, e_2, e_3) of V to the orthonormal bases $(\eta_{+1}, \eta_{+2}, \eta_{+3})$ of $\Lambda_+^2 V$ and $(\eta_{-1}, \eta_{-2}, \eta_{-3})$ of $\Lambda_-^2 V$, defined by

$$\eta_{\pm\alpha} := \tfrac{1}{\sqrt{2}}(e_0 \wedge e_\alpha \pm e_\beta \wedge e_\gamma) \qquad (2.7)$$

for each cyclic permutation (α, β, γ) of $(1, 2, 3)$. From this description also follows that $\ker \pi = \{\pm I\}$.

The action of $\mathrm{O}(V)$ on algebraic curvature tensors is now determined by the action of some orientation reversing element in $\mathrm{O}(V)$, say the one given by reversing the sign of e_0 and preserving e_1, e_2 and e_3. This element maps $\eta_{\pm\alpha}$ to $-\eta_{\mp\alpha}$. Hence its action on algebraic curvature tensors in the form (2.1) is given by conjugation with

$$\begin{pmatrix} 0 & -I \\ -I & 0 \end{pmatrix}$$

respectively by mapping

$$W_+ \mapsto W_- \qquad\qquad T_+ \mapsto T_-. \qquad (2.8)$$

2.1.3 Aligned algebraic curvature tensors

W_+ and W_- are symmetric and hence simultaneously diagonalisable under the action (2.6) of $\mathrm{SO}(V)$. On the other hand, the singular value decomposition shows that T_+ is also diagonalisable under this action, although in general not simultaneously with W_+ and W_-. The following lemma gives a criterion when this is the case.

Lemma 2.3. *Let W_+ and W_- be symmetric endomorphisms on two arbitrary Euclidean vector spaces Λ_+ respectively Λ_- and suppose the linear map $T_+ : \Lambda_+ \to \Lambda_-$ satisfies*

$$T_+ W_+ = W_- T_+.$$

Then there exist orthonormal bases for Λ_+ and Λ_- such that W_+, W_- and T_+ are simultaneously diagonal with respect to these bases.

Proof. It suffices to show that we can chose diagonal bases for W_+ and W_- such that the matrix of T_+ has at most one non-zero element in each row and in each column, for the desired result can then be obtained by an appropriate permutation of the basis elements. The above condition implies that T_+ maps eigenspaces of W_+ to eigenspaces of W_- with the same eigenvalue. Without loss of generality we can thus assume that Λ_+ and Λ_- are eigenspaces of W_+ respectively W_- with the same eigenvalue. But then W_+ and W_- are each proportional to the identity and therefore diagonal in any basis. In this case the lemma follows from the singular value decomposition for T_+. □

Lemma 2.4. *The following conditions are equivalent for an algebraic curvature tensor in the form* (2.1).

(i) $T_+ W_+ = W_- T_+$
(ii) $[\mathbf{W}, \mathbf{T}] = 0$
(iii) $\mathbf{W}$ and $\mathbf{T}$ are simultaneously diagonalisable under $SO(\Lambda^2 V)$.

Proof. The equivalence of (i) and (ii) follows from (2.3). For the equivalence of (ii) and (iii) it suffices to note that $\mathbf{W}$ and $\mathbf{T}$ are both symmetric and hence diagonalisable. □

The following terminology is borrowed from general relativity.

Definition 2.5. *We say that an algebraic curvature tensor is aligned, if it satisfies one of the equivalent conditions in Lemma 2.4.*

By Lemma 2.3, for any aligned algebraic curvature tensor we can find an orthonormal basis of V such that

$$W_\pm = \begin{pmatrix} w_{\pm 1} & 0 & 0 \\ 0 & w_{\pm 2} & 0 \\ 0 & 0 & w_{\pm 3} \end{pmatrix} \qquad T_+ = \begin{pmatrix} t_1 & 0 & 0 \\ 0 & t_2 & 0 \\ 0 & 0 & t_3 \end{pmatrix} \qquad (2.9a)$$

with

$$w_{+1} + w_{+2} + w_{+3} = 0 \qquad w_{-1} + w_{-2} + w_{-3} = 0. \qquad (2.9b)$$

To simplify notation we agree that henceforth the indices (α, β, γ) will stand for an arbitrary cyclic permutation of $(1, 2, 3)$. Changing the basis in $\Lambda^2 V$ from (2.7) to $e_i \wedge e_j$, $0 \leqslant i < j \leqslant 4$, we obtain the independent components of an aligned algebraic curvature tensor:

$$R_{0\alpha 0\alpha} = \frac{w_{+\alpha} + w_{-\alpha}}{2} + t_\alpha + \tfrac{s}{12}$$

$$R_{\beta\gamma\beta\gamma} = \frac{w_{+\alpha} + w_{-\alpha}}{2} - t_\alpha + \tfrac{s}{12} \qquad (2.10)$$

$$R_{0\alpha\beta\gamma} = \frac{w_{+\alpha} - w_{-\alpha}}{2}.$$

The trace free Ricci tensor of this aligned algebraic curvature tensor is diagonal and given by

$$\begin{aligned} T_{00} &= t_\alpha + t_\beta + t_\gamma \\ T_{\alpha\alpha} &= t_\alpha - t_\beta - t_\gamma \end{aligned} \qquad t_\alpha = \frac{T_{00} + T_{\alpha\alpha}}{2}. \qquad (2.11)$$

This is nothing else than the restriction of the isomorphism (2.4) to diagonal tensors.

2.1.4 Diagonal algebraic curvature tensors

Definition 2.6. *In view of Definition 0.12, we call an algebraic curvature tensor on V diagonalisable if it is diagonalisable as an element of $S^2\Lambda^2 V$ under the adjoint action of $\mathrm{SO}(V)$ on $\Lambda^2 V$, i.e. under the subgroup $\pi(\mathrm{SO}(V)) \subset \mathrm{SO}(\Lambda^2 V)$.*

Of course, being a symmetric form on $\Lambda^2 V$, an algebraic curvature tensor is always diagonalisable under the full group $SO(\Lambda^2 V)$.

Proposition 2.7. *An algebraic curvature tensor is diagonalisable if and only if it is aligned and W_+ has the same characteristic polynomial as W_-.*

Proof. Consider an aligned algebraic curvature tensor and suppose that W_+ has the same characteristic polynomial as W_-. Then we can find a transformation (2.6) such that, as 3×3 matrices, $W_+ = W_-$. The condition $W_- T_- = T_+ W_+$ then implies that T_+ can be simultaneously diagonalised with $W_+ = W_-$. This means we can assume without loss of generality that in (2.9) we have

$$w_{+\alpha} = w_{-\alpha} =: w_\alpha \qquad \alpha = 1, 2, 3,$$

so that (2.10) reads

$$\begin{aligned} R_{0\alpha0\alpha} &= w_\alpha + t_\alpha + \tfrac{s}{12} \\ R_{\beta\gamma\beta\gamma} &= w_\alpha - t_\alpha + \tfrac{s}{12} \end{aligned} \qquad R_{0\alpha\beta\gamma} = 0 \qquad (2.12a)$$

with w_α subject to

$$w_1 + w_2 + w_3 = 0. \qquad (2.12b)$$

Obviously this algebraic curvature tensor is diagonal. On the other hand, any diagonal algebraic curvature tensor can be written in the form (2.12), because

$$s = 2 \left(R_{0101} + R_{2323} + R_{0202} + R_{3131} + R_{0303} + R_{1212} \right) \qquad (2.13a)$$

$$t_\alpha = \frac{R_{0\alpha0\alpha} - R_{\beta\gamma\beta\gamma}}{2} \qquad (2.13b)$$

$$w_\alpha = \frac{R_{0\alpha0\alpha} + R_{\beta\gamma\beta\gamma}}{2} - \frac{s}{12}. \qquad (2.13c)$$

$\square$

2.1.5 The residual action of the isometry group

For later use we need the stabiliser of the space of diagonal algebraic curvature tensors under the action of the isometry group. Such tensors are of the form (2.9) with $w_{-\alpha} = w_{+\alpha}$ and are invariant under the transformation (2.8). Hence it is sufficient to consider only the subgroup $SO(V)$ of orientation preserving isometries.

Let $U \in SO(V)$ with $\pi(U) = (U_+, U_-)$ be an element preserving the space of diagonal algebraic curvature tensors under the action (2.6). Then $U_+ \in SO(\Lambda^2_+ V)$ and $U_- \in SO(\Lambda^2_- V)$ preserve the space of diagonal matrices on $\Lambda^2_+ V$ respectively $\Lambda^2_- V$. In the orthogonal group, the stabiliser of the space of diagonal matrices under conjugation is the subgroup of signed permutation matrices. They act by permuting the diagonal elements in disregard of the signs. In particular, in $SO(3)$ the stabiliser subgroup of the space of diagonal matrices on $\mathbb{R}^3$ is isomorphic to S_4. Under the isomorphism $S_4 \cong S_3 \ltimes V_4$, the permutation group S_3 is the subgroup of (unsigned) permutation matrices and acts by permuting the diagonal elements, whereas the Klein four group V_4 is the subgroup of diagonal matrices in $SO(3)$ and acts trivially.

Using the above, it is not difficult to see that the stabiliser of the space of diagonal algebraic curvature tensors under the action (2.6) is isomorphic to $S_4 \times \ker \pi$, with $\ker \pi = \{\pm I\}$ acting trivially. Under the isomorphism $S_4 \cong S_3 \ltimes V_4$, the factor S_3 acts by simultaneous permutations of (w_1, w_2, w_3) and (t_1, t_2, t_3) and the factor V_4 by simultaneous flips of two signs in (t_1, t_2, t_3).

2.2 Solution of the algebraic integrability conditions

2.2.1 Reformulation of the first integrability condition

In the same way as for the Bianchi identity, we can reformulate the first integrability condition (0.13a) as

$$\mathbf{R}^{ib_1}{}_{a_2 b_2} \varepsilon^{a_2 b_2}{}_{c_2 d_2} \mathbf{R}^{c_2 d_2}{}_{i d_1} = 0,$$

where $\varepsilon_{a_2 b_2 c_2 d_2}$ is the totally antisymmetric tensor, which becomes the Hodge star operator when interpreted as an endomorphims on 2-forms. Accordingly, this can be written in an index-free form as

$$r(\mathbf{R} * \mathbf{R}) = 0, \tag{2.14}$$

where

$$r: \quad \text{End}(\Lambda^2 V) \longrightarrow \text{End}(V)$$
$$\mathbf{Q}^{ij}{}_{kl} \mapsto \mathbf{Q}^{ij}{}_{kj} \tag{2.15}$$

denotes the Ricci contraction. Notice that $\mathbf{Q} := \mathbf{R} * \mathbf{R}$ is symmetric, but does in general not satisfy the Bianchi identity. A proof of the following lemma can be found in [ST69].

Lemma 2.8. *The kernel of the Ricci contraction* (2.15) *is composed of endomorphisms* $\mathbf{Q}$ *satisfying*

$$*\mathbf{Q}* = \mathbf{Q}^t \qquad\qquad \text{tr}\,\mathbf{Q} = 0.$$

Applying this to the symmetric endomorphism

$$\mathbf{Q} := \mathbf{R} * \mathbf{R} = \left(\begin{array}{c|c} (W_+ + \frac{s}{12} I_+)^2 - T_- T_+ & \begin{array}{c} (W_+ + \frac{s}{12} I_+) T_- \\ -T_-(W_- + \frac{s}{12} I_-) \end{array} \\ \hline \begin{array}{c} T_+(W_+ + \frac{s}{12} I_+) \\ -(W_- + \frac{s}{12} I_-) T_+ \end{array} & T_+ T_- - (W_- + \frac{s}{12} I_-)^2 \end{array} \right),$$

in (2.14) and using (2.1c) we get

$$T_+ W_+ = W_- T_+ \tag{2.16a}$$
$$\text{tr}\, W_+^2 = \text{tr}\, W_-^2. \tag{2.16b}$$

This shows:

Proposition 2.9. *The first algebraic integrability condition for an algebraic curvature tensor with Weyl part* $\mathbf{W}$ *and trace free Ricci part* $\mathbf{T}$ *is equivalent to*

$$[\mathbf{W}, \mathbf{T}] = 0 \qquad\qquad \text{tr}(*\mathbf{W}^2) = 0.$$

In particular, an algebraic curvature tensor satisfying the first algebraic integrability condition is aligned.

As a consequence of Proposition 2.7 we get:

Corollary 2.10. *A diagonalisable algebraic curvature tensor satisfies the first algebraic integrability condition.*

2.2.2 Integrability implies diagonalisability

The aim of this subsection is to prove part (i) of Theorem II.[1] In view of Propositions 2.7 and 2.9 it is sufficient to show that if an algebraic curvature tensor satisfies the algebraic integrability conditions, then W_+ and W_- have the same characteristic polynomial. We will first prove that the first algebraic integrability condition implies that W_+ has the same characteristic polynomial either as $+W_-$ or as $-W_-$. We then prove that the latter contradicts the second algebraic integrability condition.

We saw that an algebraic curvature tensor which satisfies the first algebraic integrability condition is aligned and thus of the form (2.9) in a suitable orthogonal basis. The first integrability condition in the form (2.16) then translates to

$$w_{+1}^2 + w_{+2}^2 + w_{+3}^2 = r^2 \qquad w_{-1}^2 + w_{-2}^2 + w_{-3}^2 = r^2 \qquad (2.17a)$$

for some $r \geqslant 0$ and

$$w_{+\alpha}t_\alpha = t_\alpha w_{-\alpha}. \qquad (2.17b)$$

If we regard (w_{+1}, w_{+2}, w_{+3}) and (w_{-1}, w_{-2}, w_{-3}) as vectors in $\mathbb{R}^3$, then each equation in (2.9b) describes the plane through the origin with normal $(1, 1, 1)$ and each equation in (2.17a) the sphere of radius r centered at the origin. Hence the solutions to both equations lie

---

[1]This result is stated without proof in [BKW85, Theorem 1] for spheres of arbitrary dimension n. Although we believe it is true in general, we include a proof here for the case $n = 3$, for the sake of self-containedness. We remark that in our algebraic setting the proof is not simple at all.

on a circle and can be parametrised in polar coordinates by a radius $r \geqslant 0$ and angles φ_+, φ_- as

$$w_{+\alpha} = r \cos\left(\varphi_+ + \alpha \tfrac{2\pi}{3}\right) \qquad w_{-\alpha} = r \cos\left(\varphi_- + \alpha \tfrac{2\pi}{3}\right). \qquad (2.18)$$

Assume now that W_+ and W_- have different eigenvalues. Then we have $T_+ = 0$. Indeed, if $t_\alpha \neq 0$ for some α, then (2.17b) shows that $w_{+\alpha} = w_{-\alpha}$ for this index α. But then (2.18) implies that W_+ and W_- have the same eigenvalues, which contradicts our assumption. Hence $T_+ = 0$.

After these considerations, the independent components (2.10) of our algebraic curvature tensor are

$$R_{0\alpha0\alpha} = \frac{w_{+\alpha} + w_{-\alpha}}{2} = R_{\beta\gamma\beta\gamma} \qquad R_{0\alpha\beta\gamma} = \frac{w_{+\alpha} - w_{-\alpha}}{2}, \qquad (2.19)$$

where we have omitted the irrelevant scalar part. Choosing $a_1 = b_1 = c_1 = d_1 = 0$ in the second integrability condition (0.13b) for this algebraic curvature tensor yields

$$g^{ij} g^{kl} \begin{array}{|c|}\hline a_2 \\\hline b_2 \\\hline c_2 \\\hline d_2 \\\hline\end{array} R_{i0a_2b_2} R_{j0k0} R_{l0c_2d_2} = \sum_{\alpha=1,2,3} \begin{array}{|c|}\hline a_2 \\\hline b_2 \\\hline c_2 \\\hline d_2 \\\hline\end{array} R_{\alpha 0 a_2 b_2} R_{\alpha 0 \alpha 0} R_{\alpha 0 c_2 d_2} = 0,$$

since only terms with $i = j = k = l \neq 0$ contribute to the contraction. For $(a_2, b_2, c_2, d_2) = (0, 1, 2, 3)$ this is gives

$$\sum R_{\alpha 0 0 \alpha} R_{\alpha 0 \alpha 0} R_{\alpha 0 \beta \gamma} = \sum (R_{0\alpha0\alpha})^2 R_{0\alpha\beta\gamma} = 0,$$

where the sums run over the three cyclic permutations (α, β, γ) of $(1, 2, 3)$. We conclude that

$$\sum_{\alpha=1,2,3} (w_{+\alpha} + w_{-\alpha})^2 (w_{+\alpha} - w_{-\alpha}) = 0.$$

Substituting (2.18) into this equation yields, after some trigonometry,

$$r^3 \cos^2 \frac{\varphi_+ - \varphi_-}{2} \sin \frac{\varphi_+ - \varphi_-}{2} \sin \frac{3}{2}(\varphi_+ + \varphi_-) = 0.$$

Substituting the solutions

$$r = 0 \qquad \varphi_+ = \varphi_- + k\pi \qquad \varphi_+ = -\varphi_- + k\tfrac{2\pi}{3} \qquad k \in \mathbb{Z}$$

of this equation back into (2.18) shows that the set $\{w_{-1}, w_{-2}, w_{-3}\}$ is either equal to $\{w_{+1}, w_{+2}, w_{+3}\}$ or to $\{-w_{+1}, -w_{+2}, -w_{+3}\}$. The first case is excluded by assumption and we conclude that W_+ and $-W_-$ have the same eigenvalues. Consequently we can choose an orthonormal basis of V such that

$$w_{+\alpha} = -w_{-\alpha} =: w_\alpha. \qquad (2.20)$$

Then the only remaining independent components of the algebraic curvature tensor (2.19) are

$$R_{0\alpha\beta\gamma} = w_\alpha. \qquad (2.21)$$

Now consider the second integrability condition (0.13b) for the indices $(a_2, b_2, c_2, d_2) = (0, 1, 2, 3)$, written as

$$\boxed{a_1\,|\,b_1\,|\,c_1\,|\,d_1}\, g^{ij} g^{kl} Q_{ib_1 l d_1} R_{ja_1 k c_1} = 0 \qquad (2.22)$$

with

$$Q_{ib_1 ld_1} := \frac{1}{4}\,\boxed{\begin{array}{c} a_2 \\ b_2 \\ c_2 \\ d_2 \end{array}}\, R_{ib_1 a_2 b_2} R_{ld_1 c_2 d_2}. \qquad (2.23)$$

The product $R_{ib_1 a_2 b_2} R_{ld_1 c_2 d_2}$ is zero unless

$$\{i, b_1, a_2, b_2\} = \{l, d_1, c_2, d_2\} = \{0, 1, 2, 3\}.$$

Since $\{a_2, b_2, c_2, d_2\} = \{0, 1, 2, 3\}$, the tensor $Q_{ib_1 ld_1}$ is zero unless

$$\{i, b_1, l, d_1\} = \{0, 1, 2, 3\}.$$

The symmetries

$$Q_{b_1 i l d_1} = -Q_{ib_1 l d_1} = Q_{ib_1 d_1 l} \qquad Q_{ib_1 l d_1} = Q_{ld_1 i b_1} \qquad (2.24)$$

then imply that the only independent components of (2.23) are

$$Q_{0\alpha\beta\gamma} = w_\alpha^2. \tag{2.25}$$

Now consider the full contraction of (2.22):

$$g^{a_1 b_1} g^{c_1 d_1} \boxed{a_1\,|\,b_1\,|\,c_1\,|\,d_1}\, Q^i{}_{b_1}{}^k{}_{d_1} R_{ia_1kc_1} = 0.$$

Since the Ricci tensor of (2.21) is zero, this is results in

$$Q^{ijkl} R_{ijkl} + Q^{ijkl} R_{ilkj} = 0.$$

Using the symmetries (2.24) of Q_{ijkl} and those of R_{ijkl}, we can always permute a particular index to the first position:

$$Q^{0jkl} R_{0jkl} + Q^{0jkl} R_{0lkj} = 0.$$

And since both Q^{ijkl} and R_{ijkl} vanish unless $\{i, j, k, l\} = \{0, 1, 2, 3\}$, we get

$$\sum (Q^{0\alpha\beta\gamma} R_{0\alpha\beta\gamma} + Q^{0\gamma\beta\alpha} R_{0\gamma\beta\alpha}) + \sum (Q^{0\alpha\beta\gamma} R_{0\gamma\beta\alpha} + Q^{0\gamma\beta\alpha} R_{0\alpha\beta\gamma}) = 0.$$

Here again, the sums run over the three cyclic permutations (α, β, γ) of $(1, 2, 3)$. Substituting (2.21) and (2.25) yields

$$\sum (w_\alpha^3 + w_\gamma^3) - \sum (w_\alpha^2 w_\gamma + w_\gamma^2 w_\alpha) = 0.$$

With (2.12b) the second term can be transformed to

$$\sum (w_\alpha^2 w_\gamma + w_\gamma^2 w_\alpha) = \sum (w_\beta^2 w_\alpha + w_\beta^2 w_\gamma) = \sum w_\beta^2 (w_\alpha + w_\gamma)$$
$$= -\sum w_\beta^3,$$

implying

$$w_1^3 + w_2^3 + w_3^3 = 0.$$

By (2.12b) and Newton's identity,

$$w_1^3 + w_2^3 + w_3^3$$
$$= 3w_1 w_2 w_3 - 3(w_1 w_2 + w_2 w_3 + w_3 w_1)(w_1 + w_2 + w_3)$$
$$+ (w_1 + w_2 + w_3)^3,$$

this is equivalent to

$$w_1 w_2 w_3 = 0.$$

Using (2.12b) once more, this shows that $\{w_1, w_2, w_3\} = \{-w, 0, +w\}$ for some $w \in \mathbb{R}$. But then (2.20) implies that $\{w_{+1}, w_{+2}, w_{+3}\} = \{w_{-1}, w_{-2}, w_{-3}\}$, which contradicts our assumption that the eigenvalues of W_+ and W_- are different. Consequently W_+ and W_- have the same eigenvalues. This finishes the proof of part (i) of Theorem II: Any integrable Killing tensor on $\mathbb{S}^3$ has a diagonalisable algebraic curvature tensor.

2.2.3 Solution of the second integrability condition

We now prove part (ii) of Theorem II. In the preceding subsections we showed that an algebraic curvature tensor which satisfies the algebraic integrability conditions is diagonal in a suitable orthogonal basis of V and that any diagonal algebraic curvature tensor satisfies the first of the two algebraic integrability conditions. We therefore now solve the second integrability condition (0.13b) for a diagonal algebraic curvature tensor in the form (2.12). To this aim consider the tensor

$$g^{ij} g^{kl} R_{ib_1 a_2 b_2} R_{j a_1 k c_1} R_{l d_1 c_2 d_2} \tag{2.26}$$

appearing on the left hand side of (0.13b). Suppose it does not vanish. Then we have $b_1 \in \{a_2, b_2\}$ and $d_1 \in \{c_2, d_2\}$, so without loss of generality we can assume $b_1 = b_2$ and $d_1 = d_2$. Then only terms with $i = j = a_2$ and $k = l = c_2$ contribute to the contraction and hence $\{a_1, a_2\} = \{c_1, c_2\}$. If now $a_2 = c_2$, then (2.26) vanishes under complete antisymmetrisation in a_2, b_2, c_2, d_2. This means that the left hand side of (0.13b) vanishes unless $\{a_1, b_1, c_1, d_1\} = \{a_2, b_2, c_2, d_2\}$. Note that a_2, b_2, c_2, d_2 and therefore a_1, b_1, c_1, d_1 have to be pairwise different due to the antisymmetrisation.

For $\dim V = 4$ in particular, (0.13b) reduces to a *sole* condition, which can be written as

$$\det \begin{pmatrix} 1 & R_{0101} + R_{2323} & R_{0101} R_{2323} \\ 1 & R_{0202} + R_{3131} & R_{0202} R_{3131} \\ 1 & R_{0303} + R_{1212} & R_{0303} R_{1212} \end{pmatrix} = 0.$$

Substituting (2.12) yields

$$\det \begin{pmatrix} 1 & w_1 & w_1^2 - t_1^2 \\ 1 & w_2 & w_2^2 - t_2^2 \\ 1 & w_3 & w_3^2 - t_3^2 \end{pmatrix} = 0. \tag{2.27}$$

This equation allows us to give an isometry invariant reformulation of the algebraic integrability conditions, which does not rely anymore on diagonalising the algebraic curvature tensor.

Proposition 2.11. *A Killing tensor on* $\mathbb{S}^3$ *is integrable if and only if its algebraic curvature tensor* $\mathbf{R}$ *satisfies*

$$[\mathbf{R}, \mathbf{R}^*] = 0 \tag{2.28a}$$
$$\chi_+ = \chi_- \tag{2.28b}$$

The identity, $\mathbf{R} + \mathbf{R}^*$ *and* $\mathbf{RR}^* = \mathbf{R}^*\mathbf{R}$ *are linearly dependent.*
$$\tag{2.28c}$$

Here χ_+ *and* χ_- *denote the characteristic polynomials of the self-dual respectively the anti-self-dual part of the Weyl tensor and* $\mathbf{R}^* := *\mathbf{R}*$ *is the Hodge dual of* $\mathbf{R}$.

Proof. This is a consequence of part (i) of Theorem II and Corollary 2.10, together with Proposition 2.7 and the second integrability condition in the form (2.27). Indeed, neglecting the scalar part, we can write $\mathbf{R} = \mathbf{W} + \mathbf{T}$, where $\mathbf{W}$ and $\mathbf{T}$ are the Weyl and the trace-free Ricci part of $\mathbf{R}$. Then $\mathbf{R}^* = \mathbf{W} - \mathbf{T}$ and Condition (2.28a) is equivalent to $[\mathbf{W}, \mathbf{T}] = 0$. Hence Condition (2.28a) is equivalent to $\mathbf{R}$ being aligned and together with Condition (2.28b) to the diagonalisability of $\mathbf{R}$. Since $\mathbf{R} + \mathbf{R}^* = 2\mathbf{W}$ and $\mathbf{RR}^* = \mathbf{W}^2 - \mathbf{T}^2$, Condition (2.28c) is the invariant form of the second integrability condition in the form (2.27). $\qquad\square$

Corollary 2.12. *A complete set of polynomial invariants for the integrability of a Killing tensor on* $\mathbb{S}^3$, *in terms of the Weyl part* $\mathbf{W}$

and the trace-free Ricci part $\mathbf{T}$ of its algebraic curvature tensor, is given by

$$\text{tr}(*\mathbf{W}^2) \quad \text{tr}[\mathbf{W},\mathbf{T}]^2 \quad t_{12} := \text{tr}(\mathbf{W}^2) \quad t_{22} := \text{tr}\big((\mathbf{W}^2 - \mathbf{T}^2)_\circ\big)^2$$
$$\text{tr}(*\mathbf{W}^3) \quad \text{tr}[\mathbf{W},\mathbf{T}]^4 \quad t_{13} := \text{tr}(\mathbf{W}^3) \quad t_{23} := \text{tr}\big((\mathbf{W}^2 - \mathbf{T}^2)_\circ\big)^3$$
$$\text{tr}[\mathbf{W},\mathbf{T}]^6 \tag{2.29}$$

where $(\mathbf{W}^2 - \mathbf{T}^2)_\circ$ denotes the trace-free part of $\mathbf{W}^2 - \mathbf{T}^2$. The Killing tensor is integrable, if and only if the invariants in the first two columns vanish and the invariants in the last two columns satisfy

$$\det \begin{pmatrix} t_{12}^3 & t_{13}^2 \\ t_{22}^3 & t_{23}^2 \end{pmatrix} = 0. \tag{2.30}$$

Proof. Together with the Bianchi identity in the form $\text{tr}(*\mathbf{W}) = 0$ the vanishing of the invariants in the first column in (2.29) is equivalent to (2.28b) and the vanishing of the invariants in the second column is equivalent to (2.28a). Since $\mathbf{W}$ is trace free, (2.28c) is equivalent to $\mathbf{W}$ and $(\mathbf{W}^2 - \mathbf{T}^2)_\circ$ being linearly dependent. This is equivalent to (2.30), as can be seen in a basis where both are diagonal. $\square$

The form (2.27) of the second algebraic integrability condition also proves part (ii) of Theorem II, since it can be rewritten in the form (0.15) by setting

$$\Delta_1 := w_2 - w_3 \quad \Delta_2 := w_3 - w_1 \quad \Delta_3 := w_1 - w_2.$$

2.3 The algebraic geometry of the Killing-Stäckel variety

We have seen that under isometries any integrable Killing tensor is equivalent to one whose algebraic curvature tensor is diagonal and that the space of these tensors is isomorphic to the following variety.

Definition 2.13. *Let* $\Sigma \cong \mathbb{P}^4$ *be the projective space of matrices whose symmetric part is diagonal and trace free, i.e. of matrices of the form*

$$M = \begin{pmatrix} \Delta_1 & -t_3 & t_2 \\ t_3 & \Delta_2 & -t_1 \\ -t_2 & t_1 & \Delta_3 \end{pmatrix} \qquad \operatorname{tr} M = 0. \qquad (2.31)$$

We denote by $\mathcal{V} \subset \Sigma$ *the Killing-Stäckel variety, i.e. the projective variety*

$$\mathcal{V} = \{M \in \Sigma : \det M = 0\}.$$

Depending on the context, we will refer to an element of $\mathcal{V}$ *as "matrix" or "point".*

Remark 2.14. *The residual isometry group action on diagonal algebraic curvature tensors defines a natural action of the permutation group* S_4 *on the Killing-Stäckel variety. This action is given by conjugation with matrices in* $\operatorname{SO}(3)$ *under the embedding* $S_4 \subset \operatorname{SO}(3)$ *defined in Subsection 2.1.5. Later it will be useful to consider* $S_4 \subset \operatorname{SO}(3)$ *as the symmetry group of an octahedron in* $\mathbb{R}^3$ *with vertices* $\pm e_i$ *and adjacent faces oppositely oriented.*

The Killing-Stäckel variety $\mathcal{V}$ is a 3-dimensional projective subvariety in $\mathbb{P}^4$ and the quotient $\mathcal{V}/S_4$ encodes all information on integrable Killing tensors modulo isometries and affine transformations. In this section we will investigate its structure from a purely algebraic geometric point of view. This will finally lead to a complete and explicit algebraic description of Stäckel systems.

Recall that Stäckel systems on $\mathbb{S}^3$ are 3-dimensional vector spaces of mutually commuting integrable Killing tensors, that every Stäckel system contains the metric and that the metric corresponds to the zero matrix in $\mathcal{V}$. Therefore, if we assume that the algebraic curvature tensors of the Killing tensors in a Stäckel system are mutually diagonalisable, then Stäckel systems on $\mathbb{S}^3$ correspond to projective lines in $\mathcal{V}$. We will see a posteriori that our assumption is justified. Hence

Stäckel systems constitute a subvariety of the variety of projective lines on $\mathcal{V}$, also called the Fano variety of $\mathcal{V}$ and denoted by $F_1(\mathcal{V})$.

Let $\mathcal{D}$ be the generic determinantal variety, i.e. the projective variety of 3×3 matrices with vanishing determinant or, equivalently, 3×3 matrices of rank one or two. By definition, $\mathcal{V} = \mathcal{D} \cap \Sigma$ is a subvariety of $\mathcal{D}$. Hence its Fano variety $F_1(\mathcal{V})$ is a subvariety of the Fano variety $F_1(\mathcal{D})$. But the latter is well understood. It contains three different types of projective spaces:

(i) projective spaces of matrices whose kernel contains a fixed line,
(ii) projective spaces of matrices whose image is contained in a fixed plane and
(iii) the projective plane of antisymmetric matrices.

This motivates why we seek for such spaces in $\mathcal{V}$. Note that the projective plane of antisymmetric matrices is obviously contained in $\mathcal{V}$.

Definition 2.15. *We call a projective subspace in the KS variety consisting of matrices whose kernel contains a fixed line (respectively whose image is contained in a fixed plane) an* isokernel *space (respectively an* isoimage *space).*

A 3×3 matrix M of rank two has a 2-dimensional image and a 1-dimensional kernel. Both are given by its adjugate matrix $\mathrm{Adj}\, M$, which is the transpose of the cofactor matrix and satisfies

$$(\mathrm{Adj}\, M)M = M(\mathrm{Adj}\, M) = (\det M)I. \qquad (2.32)$$

The adjugate of a matrix $M \in \mathcal{V}$ is

$$\mathrm{Adj}\, M = \begin{pmatrix} t_1 t_1 + \Delta_2 \Delta_3 & t_2 t_1 + \Delta_3 t_3 & t_3 t_1 - \Delta_2 t_2 \\ t_1 t_2 - \Delta_3 t_3 & t_2 t_2 + \Delta_3 \Delta_1 & t_3 t_2 + \Delta_1 t_1 \\ t_1 t_3 + \Delta_2 t_2 & t_2 t_3 - \Delta_1 t_1 & t_3 t_3 + \Delta_1 \Delta_2 \end{pmatrix}. \qquad (2.33)$$

The adjugate matrix of a rank two 3×3 matrix has rank one, which means that the columns of $\mathrm{Adj}\, M$ span a line in $\mathbb{P}^2$. And because

$\det M = 0$, we deduce from (2.32) that the kernel of a rank two matrix M is the column space of $\mathrm{Adj}\, M$.

The adjugate matrix of a rank one 3×3 matrix is zero. It is known, that the singular locus of $\mathcal{D}$ is the subvariety of rank one matrices. The following proposition characterises the singular locus of $\mathcal{V} = \mathcal{D} \cap \Sigma$.

Proposition 2.16. *The KS variety $\mathcal{V}$ has ten singular points, given by the six rank one matrices*

$$
V_{+1} := \begin{pmatrix} 0 & 0 & 0 \\ 0 & +1 & -1 \\ 0 & +1 & -1 \end{pmatrix} \qquad V_{-1} := \begin{pmatrix} 0 & 0 & 0 \\ 0 & +1 & +1 \\ 0 & -1 & -1 \end{pmatrix}
$$

$$
V_{+2} := \begin{pmatrix} -1 & 0 & +1 \\ 0 & 0 & 0 \\ -1 & 0 & +1 \end{pmatrix} \qquad V_{-2} := \begin{pmatrix} -1 & 0 & -1 \\ 0 & 0 & 0 \\ +1 & 0 & +1 \end{pmatrix} \qquad (2.34a)
$$

$$
V_{+3} := \begin{pmatrix} +1 & -1 & 0 \\ +1 & -1 & 0 \\ 0 & 0 & 0 \end{pmatrix} \qquad V_{-3} := \begin{pmatrix} +1 & +1 & 0 \\ -1 & -1 & 0 \\ 0 & 0 & 0 \end{pmatrix}
$$

and the four skew symmetric matrices

$$
C_0 := V_{+1} + V_{+2} + V_{+3} \qquad C_\alpha := V_{-\alpha} + V_{+\beta} + V_{+\gamma}, \qquad (2.34b)
$$

where (α, β, γ) denotes any cyclic permutation of $(1, 2, 3)$. Moreover, the rank one matrices in $\mathcal{V}$ are exactly those six in (2.34a).

Definition 2.17. *We will call the six singularities $V_{\pm\alpha}$ the rank one singular points in $\mathcal{V}$ and the four singularities C_i the skew symmetric singular points in $\mathcal{V}$.*

Proof. Let $\hat{\Sigma}$ be the linear space of matrices of the form (0.15a), so that $\mathcal{V} = \mathbb{P}\det^{-1}(0)$ is the projectivisation of the zero locus of the determinant $\det \colon \hat{\Sigma} \to \mathbb{R}$. Then $\mathcal{V}$ is singular at M if and only if the derivative of the determinant at M, given by

$$
(d \det)_M \colon \begin{aligned} \hat{\Sigma} &\longrightarrow \mathbb{R} \\ A &\mapsto \mathrm{tr}(A \,\mathrm{Adj}\, M), \end{aligned} \qquad (2.35)
$$

is the zero map. The condition that $\text{tr}(A \, \text{Adj} \, M) = 0$ for all $A \in \Sigma$ is equivalent to

$$t_1^2 + \Delta_2\Delta_3 = t_2^2 + \Delta_3\Delta_1 = t_3^2 + \Delta_1\Delta_2 \quad \Delta_1 t_1 = \Delta_2 t_2 = \Delta_3 t_3 = 0.$$

We leave it to the reader to verify that the solutions $M \in \mathcal{V}$ of these equations are exactly the matrices (2.34) and their multiples.

The last statement now follows from (2.35) and the equivalence of $\text{rank} \, M = 1$ and $\text{Adj} \, M = 0$ for 3×3-matrices. $\qquad\square$

Remark 2.18. *Under the natural S_4-action on $\mathcal{V}$, all rank one singular points are equivalent. This means that the singularities $\mathcal{V}_{\pm\alpha}$ are all mapped to a single point in the quotient $\mathcal{V}/S_4$. The same holds for the skew symmetric singular points.*

We now compute the isokernel spaces in $\mathcal{V}$. To find all matrices $M \in \mathcal{V}$ annihilating a given vector $\vec{n} \in \mathbb{R}^3$, we consider the equation $M\vec{n} = 0$ as an equation in M for a fixed $\vec{n} = (n_1, n_2, n_3)$, where M is of the form $M = \Delta + T$ with Δ the diagonal part and T the skew symmetric part. We can regard $\vec{t} = (t_1, t_2, t_3) \in \mathbb{R}^3$ as a parameter and solve $\Delta\vec{n} = -T\vec{n}$ for Δ. For the time being, let us assume $n_1 n_2 n_3 \neq 0$. We then obtain a linear family of matrices

$$M = \begin{pmatrix} \frac{n_2}{n_1}t_3 - \frac{n_3}{n_1}t_2 & -t_3 & t_2 \\ t_3 & \frac{n_3}{n_2}t_1 - \frac{n_1}{n_2}t_3 & -t_1 \\ -t_2 & t_1 & \frac{n_1}{n_3}t_2 - \frac{n_2}{n_3}t_1 \end{pmatrix}. \tag{2.36a}$$

All these matrices satisfy $\det M = 0$ and the condition $\text{tr} \, M = 0$ imposes the restriction

$$\vec{t} \perp \left(\frac{n_3}{n_2} - \frac{n_2}{n_3}, \frac{n_1}{n_3} - \frac{n_3}{n_1}, \frac{n_2}{n_1} - \frac{n_1}{n_2} \right) \tag{2.36b}$$

on the parameter $\vec{t} \in \mathbb{R}^3$. This shows that the matrices in $\mathcal{V}$ annihilating a fixed vector $\vec{n} \neq 0$ form a projective subspace. If the right hand side of (2.36b) is zero, i.e. if the condition is void, this subspace has dimension three and defines a projective plane in $\mathcal{V}$. This happens if and only if

$$|n_1| = |n_2| = |n_3|. \tag{2.37}$$

If the right hand side of (2.36b) is not zero, this subspace has dimension two and defines a projective line in $\mathcal{V}$. One can check that this also holds true for the case $n_1 n_2 n_3 = 0$. We have shown the following:

Lemma 2.19. *If $n \in \mathbb{P}^2$ satisfies (2.37), then the set of matrices in $\mathcal{V}$ annihilating $\vec{n}$ is a projective plane of the form (2.36a). Otherwise it is a projective line.*

Recall that the kernel of a rank two matrix of the form (2.31) is the row space of the adjugate matrix (2.33). This defines a rational map

$$\pi \colon \mathcal{V} \dashrightarrow \mathbb{P}^2$$

whose fibres are the maximal isokernel spaces. This map is well defined except for the six rank one matrices in $\mathcal{V}$.

We want to give a parametrisation of the isokernel lines that is uniform and more geometric than (2.36). For this purpose we define two embeddings

$$\iota, \nu \colon \mathbb{P}^2 \to \mathcal{V}$$

given by

$$
\iota(n) = \begin{pmatrix} 0 & -n_3 & n_2 \\ n_3 & 0 & -n_1 \\ -n_2 & n_1 & 0 \end{pmatrix}
$$

$$
\nu(n) = \begin{pmatrix} n_2^2 - n_3^2 & -n_1 n_2 & n_3 n_1 \\ n_1 n_2 & n_3^2 - n_1^2 & -n_2 n_3 \\ -n_3 n_1 & n_2 n_3 & n_1^2 - n_2^2 \end{pmatrix} \tag{2.38}
$$

for $n = (n_1 : n_2 : n_3) \in \mathbb{P}^2$. One readily checks that $\iota(n)$ and $\nu(n)$ have trace and determinant equal to zero and thus indeed lie in $\mathcal{V}$. Note that the image of ι is the projective plane of skew symmetric matrices in $\mathcal{V}$. Under the inclusion $\mathcal{V} \subset \mathbb{P}\Sigma \cong \mathbb{P}^4$ the map ι is a linear embedding of $\mathbb{P}^2$ in $\mathbb{P}^4$ whilst ν is the composition of the Veronese embedding

$$\mathbb{P}^2 \hookrightarrow \mathbb{P}^5$$
$$(n_1 : n_2 : n_3) \mapsto \left(n_1^2 : n_2^2 : n_3^2 : n_2 n_3 : n_3 n_1 : n_1 n_2 \right)$$

and a projection $\mathbb{P}^5 \twoheadrightarrow \mathbb{P}^4$ under which the Veronese surface remains smooth.

Proposition 2.20 (Isokernel spaces in the KS variety). *Recall from Proposition 2.16 that the singular points in the KS variety $\mathcal{V}$ are the four skew symmetric singular points C_0, C_1, C_2, C_3 as well as six rank one singular points $V_{\pm 1}, V_{\pm 2}, V_{\pm 3}$, and let the maps $\pi \colon \mathcal{V} \dashrightarrow \mathbb{P}^2$ and $\iota, \nu \colon \mathbb{P}^2 \to \mathcal{V}$ be defined as above.*

(i) The projective plane through V_{+1}, V_{+2} and V_{+3} is an isokernel plane in $\mathcal{V}$ and contains C_0. In the same way, the three points $V_{+\alpha}$, $V_{-\beta}$ and $V_{-\gamma}$ define an isokernel plane through C_α for (α, β, γ) a cyclic permutation of $(1, 2, 3)$.

(ii) For any non-singular point $M \in \mathcal{V}$ the following points are collinear, but do not all three coincide:

$$M \qquad \iota(\pi(M)) \qquad \nu(\pi(M))$$

The projective line they define lies in $\mathcal{V}$ and is an isokernel space through M. Each of these lines contains a unique skew symmetric point, namely $\iota(\pi(M))$.

Moreover, any isokernel space in $\mathcal{V}$ of maximal dimension is of either of the above forms and hence contains a unique skew symmetric point.

Remark 2.21. *A corresponding characterisation of isoimage spaces follows from the fact that matrix transposition defines an involution of $\mathcal{V}$ which interchanges isokernel and isoimage spaces.*

Proof. (i) follows directly from Lemma 2.19 and the explicit form (2.34) of the singular points. For (ii) notice that $n := \pi(M)$ is the kernel of M and well defined for $M \neq V_\alpha$. The definitions (2.38) of ι and ν then show that n is also the kernel of $\iota(n)$ and $\nu(n)$. These definitions also show that M, $\iota(n)$ and $\nu(n)$ do not all coincide for $M \neq C_i$. Their collinearity follows from Lemma 2.19 in the case where n does not satisfy (2.37) and from $\iota(n) = \nu(n)$ in case it does. The last statement is now a consequence of Lemma 2.19. $\qquad\square$

We can illustrate the content of Proposition 2.20 geometrically as follows. This is depicted in Figure 2.1. The six rank one singular points of $\mathcal{V} \subset \mathbb{P}^4$ constitute the vertices of a regular octahedron in $\mathbb{P}^4$ whose faces (and the planes they define) are entirely contained in $\mathcal{V}$. The set of eight faces is divided into two sets of non-adjacent faces, corresponding to the four isokernel planes (shaded) respectively the four isoimage planes (not shaded). Opposite faces intersect in their respective centres. These are the skew symmetric singular points of $\mathcal{V}$ and the intersection points of the octahedron with the projective plane of skew symmetric matrices. In Figure 2.1 this ninth projective plane in $\mathcal{V}$ is depicted as the insphere of the octahedron.

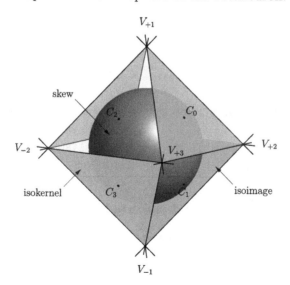

Figure 2.1: Singularities and projective planes in the KS variety.

The permutation group S_4 acts on this configuration by symmetries of the octahedron preserving the two sets of non-adjacent faces. Matrix transposition corresponds to a point reflection, which exchanges opposite faces and completes the S_4-action to the full octahedral symmetry.

2.4 Interpretation of the Killing-Stäckel variety

In this section we set out a correspondence between the algebro-geometric properties of the Killing-Stäckel variety and the geometric properties of the corresponding Killing tensors. This is summarised in Table 2.1 and the details will be made clear in the following.

Table 2.1: Correspondence between algebro-geometric properties of the Killing-Stäckel variety and geometric properties of Killing tensors

algebro-geometric properties of the KS variety	geometric properties of integrable Killing tensors
singularities	metrics on embedded spheres
– rank one singular points	– metrics of embedded circles
– skew symmetric singular points	– metrics of embedded 2-spheres
projective plane of antisymmetric matrices	special Killing tensors
	Stäckel systems
certain isokernel lines	Stäckel systems of embedded
isokernel planes	2-spheres

2.4.1 Stäckel systems and isokernel lines

In this subsection we identify the Stäckel systems as projective lines in the KS variety.

Definition 2.22. *We call the isokernel lines of the form (ii) in Proposition 2.20 Stäckel lines.*

By definition, every Stäckel line intersects the projective plane of skew symmetric matrices transversely. Conversely, every point on this plane determines a unique Stäckel line unless it is one of the four singular skew symmetric points C_i. The Stäckel lines through these points are exactly those projective lines in each isokernel plane which pass through the respective face center C_i.

Proposition 2.23. *Stäckel lines in the KS variety correspond to Stäckel systems. More precisely, the preimage of a Stäckel line under the quotient*

$$\mathcal{I}(\mathbb{S}^3) \cap \mathcal{K}_0(\mathbb{S}^3) \longrightarrow \left(\mathcal{I}(\mathbb{S}^3) \cap \mathcal{K}_0(\mathbb{S}^3)\right)/\mathrm{Aff}(\mathbb{R}) \cong \mathcal{V} \qquad (2.39)$$

is a Stäckel system.

Proof. A Stäckel line is the projective line through the (collinear but not coinciding) points M, $\iota(\pi(M))$ and $\nu(\pi(M))$ in the KS variety for any non-singular point $M \in \mathcal{V}$. We have to prove that any preimages K_0, K_1 and K_2 under the quotient (2.39) mutually commute. We first show that K_1 and K_2 commute. Let R and $\tilde{R}$ be their diagonal algebraic curvature tensors. From their matrices (2.38) and $\Delta_\alpha = w_\beta - w_\gamma$ we can read off the values

$$w_\alpha = 0 \qquad t_\alpha = n_\alpha \qquad \tilde{w}_\alpha = n_\alpha^2 \qquad \tilde{t}_\alpha = n_\beta n_\gamma,$$

where we have absorbed the scalar curvature into the w_α's by abandoning the restriction that they sum to zero. The diagonals of R and $\tilde{R}$ are then given by the columns of the matrix

$$
\begin{pmatrix}
1 & R_{0101} & \tilde{R}_{0101} \\
1 & R_{0202} & \tilde{R}_{0202} \\
1 & R_{0303} & \tilde{R}_{0303} \\
1 & R_{2323} & \tilde{R}_{2323} \\
1 & R_{3131} & \tilde{R}_{3131} \\
1 & R_{1212} & \tilde{R}_{1212}
\end{pmatrix}
=
\begin{pmatrix}
1 & w_1+t_1 & \tilde{w}_1+\tilde{t}_1 \\
1 & w_2+t_2 & \tilde{w}_2+\tilde{t}_2 \\
1 & w_3+t_3 & \tilde{w}_3+\tilde{t}_3 \\
1 & w_1-t_1 & \tilde{w}_1-\tilde{t}_1 \\
1 & w_2-t_2 & \tilde{w}_2-\tilde{t}_2 \\
1 & w_3-t_3 & \tilde{w}_3-\tilde{t}_3
\end{pmatrix}
=
\begin{pmatrix}
1 & +n_1 & n_1^2+n_2 n_3 \\
1 & +n_2 & n_2^2+n_3 n_1 \\
1 & +n_3 & n_3^2+n_1 n_2 \\
1 & -n_1 & n_1^2-n_2 n_3 \\
1 & -n_2 & n_2^2-n_3 n_1 \\
1 & -n_3 & n_3^2-n_1 n_2
\end{pmatrix}.
$$

The four square matrices in (1.44) can be obtained from this matrix by discarding the rows $(1,2,3)$, $(1,5,6)$, $(2,4,6)$ respectively $(3,4,5)$ and it is not difficult to see that their determinants vanish. Corollary 1.18 therefore shows that K_1 and K_2 commute.

If $\iota(\pi(M))$ and $\nu(\pi(M))$ are linearly independent, then M is a linear combination of both and hence K_0 is a linear combination of g, K_1 and K_2. In other words, K_0, K_1 and K_2 mutually commute.

If $\iota(\pi(M))$ and $\nu(\pi(M))$ are linearly dependent, then $n = \pi(M)$ satisfies $|n_1| = |n_2| = |n_3|$. For simplicity, suppose $n_1 = n_2 = n_3 = 1$. By Lemma 2.19, M is of the form (2.36a) and thus has $w_\alpha = -t_\alpha$. Consequently, the diagonal algebraic curvature tensors R and $\tilde{R}$ of K_0 respectively K_1 are given by

$$
\begin{pmatrix}
1 & R_{0101} & \tilde{R}_{0101} \\
1 & R_{0202} & \tilde{R}_{0202} \\
1 & R_{0303} & \tilde{R}_{0303} \\
1 & R_{2323} & \tilde{R}_{2323} \\
1 & R_{3131} & \tilde{R}_{3131} \\
1 & R_{1212} & \tilde{R}_{1212}
\end{pmatrix}
=
\begin{pmatrix}
1 & w_1+t_1 & \tilde{w}_1+\tilde{t}_1 \\
1 & w_2+t_2 & \tilde{w}_2+\tilde{t}_2 \\
1 & w_3+t_3 & \tilde{w}_3+\tilde{t}_3 \\
1 & w_1-t_1 & \tilde{w}_1-\tilde{t}_1 \\
1 & w_2-t_2 & \tilde{w}_2-\tilde{t}_2 \\
1 & w_3-t_3 & \tilde{w}_3-\tilde{t}_3
\end{pmatrix}
=
\begin{pmatrix}
1 & 0 & +1 \\
1 & 0 & +1 \\
1 & 0 & +1 \\
1 & 2t_1 & -1 \\
1 & 2t_2 & -1 \\
1 & 2t_3 & -1
\end{pmatrix}.
$$

As above, they satisfy the commutation condition (1.44) and we conclude that K_0, K_1 and K_2 commute. $\qquad\square$

Corollary 2.24. *The Killing tensors in a Stäckel system on $\mathbb{S}^3$ have simultaneously diagonalisable algebraic curvature tensors.*[2]

Proof. This follows from the fact that every Stäckel system contains a Killing tensor with simple eigenvalues and that such a Killing tensor K uniquely determines the Stäckel system [Ben93]. By the very definition of a Stäckel system, K is integrable and so we can assume it to have a diagonal algebraic curvature tensor. The corresponding point on the KS variety lies on a Stäckel line. By the above, K then lies in a Stäckel system with diagonal algebraic curvature tensors. This proves the statement, since the Stäckel system determined by K is unique. $\qquad\square$

---

[2]Footnote 1 on page 66 applies here as well.

2.4.2 Antisymmetric matrices and special conformal Killing tensors

In this subsection we give an interpretation of the projective plane of antisymmetric matrices in the KS variety. They will be closely related to special conformal Killing tensors (cf. Definition 0.11).

Lemma 2.25. *If L is a special conformal Killing tensor, then*

$$K := L - (\operatorname{tr} L)g \tag{2.40}$$

is a Killing tensor. This defines an injective linear map from the space of special conformal Killing tensors to the space of Killing tensors.

Proof. It is straightforward to check the first statement. Taking the trace on both sides of $K = L - (\operatorname{tr} L)g$ yields $\operatorname{tr} K = (1 - n) \operatorname{tr} L$ and hence $L = K - \frac{\operatorname{tr} K}{n-1}g$. This shows injectivity. □

Definition 2.26. *We define a special Killing tensor to be a Killing tensor of the form* (2.40), *where L is a special conformal Killing tensor.*

Lemma 2.27. *A special conformal Killing tensor as well as the corresponding special Killing tensor have vanishing Nijenhuis torsion. In particular, both are integrable.*

Proof. Substituting Equation (0.10) into the expression (0.2) for $K = L$ shows that the Nijenhuis torsion of L is zero. Substituting $K = L - (\operatorname{tr} L)g$ into (0.2) shows that this implies that the Nijenhuis torsion of K is also zero. □

As with Killing tensors, we can characterise special conformal Killing tensors on constant curvature manifolds algebraically. This is a special case of a fact which holds under more general assumptions than constant curvature, namely that special conformal Killing tensors extend to covariantly constant tensors on the metric cone (see e.g. [MM10]).

Lemma 2.28. *The space of special conformal Killing tensors on a constant curvature manifold $M \subset V$ is isomorphic to the space of of (constant) symmetric forms $\hat{L}$ on V. The isomorphism is given by restricting $\hat{L}$ to M.*

Proof. We leave it to the reader to check that such a restriction satisfies the defining equation (0.10) of a special conformal Killing tensor. This shows that restriction yields a well defined map from symmetric tensors on V to special conformal Killing tensors on $M \subset V$. The injectivity of this map is evident and surjectivity follows from dimension considerations, since the dimension of the space of conformal Killing tensors on an n-dimensional constant curvature manifold is known to be $\binom{n+1}{2}$ [Cra03]. $\qquad\square$

Corollary 2.29. *A Killing tensor on a constant curvature manifold is special if and only if its Weyl tensor is zero.*

Proof. By Lemma 2.25 the space of special Killing tensors is a subrepresentation of the space of Killing tensors. By Lemma 2.28 it has dimension $\binom{n+1}{2}$. The statement now follows from the observation that the only subrepresentation of the space of algebraic curvature tensors on an $(n + 1)$-dimensional space having this dimension is the Ricci part, i.e. the subspace of algebraic curvature tensors with vanishing Weyl part. $\qquad\square$

Remark 2.30. *The above lemma can also be proven directly by showing that the isomorphism from the space of special conformal Killing tensors to the space of special Killing tensors, given by (2.40), is given algebraically by mapping the symmetric form $\hat{L}$ to the algebraic curvature tensor* [3]

$$R_{a_1 b_1 a_2 b_2} = \hat{L}_{a_1 a_2} g_{b_1 b_2} - \hat{L}_{a_1 b_2} g_{b_1 a_2} - \hat{L}_{b_1 a_2} g_{a_1 b_2} + \hat{L}_{b_1 b_2} g_{a_1 a_2} \quad (2.41)$$

(up to an irrelevant trace term). In other words, up to an affine transformation $\hat{L}$ is the Ricci tensor of the corresponding special Killing tensor.

---

[3]Expressed with the Kulkarni-Nomizu product (2.44), this reads $R = L \oslash g$.

Corollary 2.31. *The space of special Killing tensors on $\mathbb{S}^3$ corresponds to the projective space of antisymmetric matrices inside the KS variety.*

Proof. In an orthogonal basis where the symmetric form $\hat{L}$ is diagonal with diagonal entries Λ_i, (2.41) is clearly diagonal with diagonal entries $R_{ijij} = \Lambda_i + \Lambda_j$. From (2.13) we conclude that

$$w_\alpha = 0 \qquad\qquad t_\alpha = \Lambda_0 + \Lambda_\alpha. \qquad (2.42)$$

and hence $\Delta_\alpha = 0$. This proves that the matrix (0.15a) of a special Killing tensor is antisymmetric. The converse follows from dimension considerations, since the space of diagonal Ricci tensors modulo affine transformations as well as the projective space of antisymmetric 3×3 matrices have both dimension two. □

The following is now a consequence of Propositions 2.20 and 2.23.

Corollary 2.32. *Every Stäckel system on $\mathbb{S}^3$ contains a special Killing tensor which is unique up to affine transformations.*

In the next section we will make use of this essentially unique representative in each Stäckel system in order to derive the corresponding separation coordinates.

2.4.3 Isokernel planes and integrable Killing tensors from $\mathbb{S}^2$

In this subsection we give an interpretation of the isokernel planes in the KS variety. They will be Killing tensors coming from $\mathbb{S}^2$ in the following sense. A Killing tensor on an isometrically embedded sphere $\mathbb{S}^{n-1} \subset \mathbb{S}^n$, given as the intersection of $\mathbb{S}^n \subset V$ with a hyperplane $H \subset V$, can be extended to the ambient sphere $\mathbb{S}^n$ by extending the corresponding algebraic curvature tensor R on H by zero to the ambient space V, i.e. by defining the extension $\hat{R}(w, x, y, z)$ on V to be equal to $R(w, x, y, z)$ if $w, x, y, z \in H$ and to be zero whenever one of the arguments lies in $H^\perp$. This defines an inclusion $\mathcal{K}(\mathbb{S}^{n-1}) \hookrightarrow \mathcal{K}(\mathbb{S}^n)$ for each hyperplane $H \subset V$.

Moreover, this extension preserves integrability, as can be seen, for example, from the algebraic integrability conditions (0.13). In other words, for each hyperplane $H \subset V$ the inclusion of subspaces $\mathcal{K}(\mathbb{S}^{n-1}) \hookrightarrow \mathcal{K}(\mathbb{S}^n)$ restricts to an inclusion $\mathcal{I}(\mathbb{S}^{n-1}) \hookrightarrow \mathcal{I}(\mathbb{S}^n)$ of varieties. Restricting to the space of diagonal algebraic curvature tensors $\mathcal{K}_0(\mathbb{S}^n) \subset \mathcal{K}(\mathbb{S}^n)$ means to fix a basis of the $(n+1)$-dimensional vector space V and to choose a subset of this basis as a basis for H. We have $n + 1$ choices for H. This is why we will find $n + 1$ copies of the variety $\mathbb{P}(\mathcal{I}(\mathbb{S}^{n-1}) \cap \mathcal{K}_0(\mathbb{S}^{n-1}))$ as subvarieties of the KS variety $\mathcal{V} \cong (\mathcal{I}(\mathbb{S}^n) \cap \mathcal{K}_0(\mathbb{S}^n))/\mathrm{Aff}(\mathbb{R})$, all copies being equivalent under isometries. Note that while the metric on $\mathbb{S}^n$ does not appear in $\mathcal{V}$, because we factor by affine transformations, $\mathcal{V}$ does contain copies of the metric from $\mathbb{S}^{n-1}$. The reason is that the extension of Killing tensors is not compatible with the affine action. In particular, by virtue of the examples in Section 0.7.3 we will find four projective planes of integrable Killing tensors from $\mathbb{S}^2$ inside the KS variety. The Stäckel systems they contain all meet in the metric from $\mathbb{S}^2$. This explains the following.

Proposition 2.33. *The isokernel planes in the KS variety consist of Killing tensors extended from $\mathbb{S}^2 \subset \mathbb{S}^3$ in the sense explained above. In particular, the singularities of the KS variety correspond to the metric on embedded spheres: The four skew symmetric singular points correspond to the metrics of embedded 2-spheres and the six rank one singular points to the metrics of embedded circles.*

2.5 Separation coordinates

We eventually demonstrate how the orthogonal separation coordinates on $\mathbb{S}^3$ and their classification can be derived from our algebraic description. Although this construction will be superseded by the one in Section 3.7.2, we need its relation to the KS variety in order to derive polynomial invariants for the classification of separation coordinates.

In Corollary 2.32 we have proven that every Stäckel system on $\mathbb{S}^3$ contains a special Killing tensor which is unique up to affine transformations. We have also seen that the Stäckel system is uniquely defined by this special Killing tensor unless it comes from a lower dimensional sphere. By definition, a special Killing tensor $K = L - (\operatorname{tr} L)g$ has the same eigenspaces as its corresponding special conformal Killing tensor L. Moreover, the eigenvalues of a special conformal Killing tensor are constant on the hypersurfaces orthogonal to the corresponding eigenvectors [Cra03]. That is to say we can actually take the eigenvalues of the special conformal Killing tensor in the Stäckel system as coordinate functions whenever these eigenvalues are simple and non-constant. Thereby the classification of orthogonal separation coordinates on $\mathbb{S}^3$ is reduced to a computation of the eigenvalues of special conformal Killing tensors. This computation is considerably simplified by the fact that a special conformal Killing tensor on a constant curvature manifold is the restriction of a (constant) symmetric form (cf. Lemma 2.28).

Let L be a special conformal Killing tensor on a non-flat constant curvature manifold $M \subset V$, i.e. the restriction from V to M of a (constant) symmetric form $\hat{L}$. From now on we consider L and $\hat{L}$ as endomorphisms. We want to compute the eigenvalues $\lambda_1(x), \ldots, \lambda_n(x)$ of $L \colon T_x M \to T_x M$ for a fixed point $x \in M$ and relate them to the (constant) eigenvalues $\Lambda_0, \ldots, \Lambda_n$ of $\hat{L} \colon V \to V$.

Let $P : V \to V$ be the orthogonal projection from $V \cong T_x M \oplus \mathbb{R}x$ to $T_x M$. Then the restriction of $P\hat{L}P$ to $T_x M$ is L and the restriction of $P\hat{L}P$ to x is zero. Therefore x is an eigenvector of $P\hat{L}P$ with eigenvalue 0 and the remaining eigenvectors and eigenvalues are those of L. This means the eigenvalue equation $Lv = \lambda v$ is equivalent to $P\hat{L}Pv = \lambda v$ for $v \perp x$. By the definition of P we have $Pv = v$ and $P\hat{L}v = \hat{L}v - g(\hat{L}v, x)x$. Hence we seek common solutions to the two equations

$$(\hat{L} - \lambda)v = g(\hat{L}v, x)x \qquad\qquad g(x, v) = 0.$$

If λ is not an eigenvalue of $\hat{L}$, then $(\hat{L} - \lambda)$ is invertible and v is proportional to $(\hat{L} - \lambda)^{-1}x$. The condition $v \perp x$ then yields the equation

$$g\big((\hat{L} - \lambda)^{-1}x, x\big) = 0$$

for λ. In an eigenbasis of $\hat{L}$ this equation is nothing but the defining equation (0.11) for elliptic coordinates. The solutions of this equation are the eigenvalues of L and can be ordered according to

$$\Lambda_0 \leqslant \lambda_1(x) \leqslant \Lambda_1 \leqslant \lambda_2(x) \leqslant \cdots \leqslant \lambda_n(x) \leqslant \Lambda_n.$$

If $\hat{L}$ has simple eigenvalues, then the eigenvalues of L are also simple and define elliptic coordinates.

If $\hat{L}$ has a double eigenvalue $\Lambda_{k-1} = \Lambda_k$, then the eigenvalues of L are still simple, but the corresponding eigenvalue λ_k of L is constant and can therefore not be used as a coordinate. Nevertheless, we can recover the missing coordinate function from the fact that the coordinate hypersurfaces are orthogonal to the eigenspaces of L. Let E be an eigenplane of $\hat{L}$ and denote by $E^\perp$ its orthogonal complement in V. Then the corresponding eigenspace of L at a point $x \in M$ is given by the intersection of E with the tangent space $T_x M$. Its orthogonal complement in $T_x M$ is the intersection of $T_x M$ with the hyperplane that contains both x and $E^\perp$. The coordinate hypersurfaces are therefore the intersections of hyperplanes containing $E^\perp$ with M and can be parametrised by the polar angle in E. That is, we can replace the constant eigenvalue by an angular coordinate.

If $\hat{L}$ has eigenvalues of multiplicity greater than two, the situation becomes more involved and will be elucidated in Section 3.7.2. However, at this point we can already tell what happens for $\mathbb{S}^3 \subset \mathbb{R}^4$, as in this case $\hat{L}$ can only have one triple and one simple eigenvalue.[4] Under affine transformations the associated special Killing tensor K is equivalent to the extension of the metric on $\mathbb{S}^2$. Hence K is contained in all Stäckel systems which are extensions from $\mathbb{S}^2$. This simply means that K is not generic in such a Stäckel system and we have to consider another representative in order to derive the separation coordinates. Up to an affine transformation, we can choose this representative to be an extension from $\mathbb{S}^2$. That is, we have to go down one dimension and to repeat the same procedure there. On $\mathbb{S}^2 \subset \mathbb{R}^3$ there are only two possible multiplicities for $\hat{L}$: either only simple

---

[4]All eigenvalues equal would mean that $\hat{L}$ is trivial under affine transformations.

Table 2.2: Classification of separation coordinates on $\mathbb{S}^3$.

class	coordinates	induced from	[Eis34]	[KM86]
(0123)	elliptic	–	V	(1)
(01(23))	oblate Lamé rotational	–	I	(2a)
(0(12)3)	prolate Lamé rotational	–	I	(2b)
((01)(23))	cylindrical	$\mathbb{S}^1 \times \mathbb{S}^1$	III	(5)
(0(123))	Lamé subgroup reduction	$\mathbb{S}^2$ elliptic	IV	(3)
(0(1(23)))	spherical	$\mathbb{S}^2$ spherical	II	(4)

eigenvalues or one simple and one double eigenvalue, corresponding to elliptic and spherical coordinates respectively.

To summarise the above, the multiplicities of the eigenvalues of the symmetric form $\hat{L}$ characterise the different classes of separation coordinates, provided it has at most double eigenvalues. Otherwise they are extensions from $\mathbb{S}^2$ and we have to descend recursively. Consequently the different families of separation coordinates on $\mathbb{S}^3$ can be labelled by nested parentheses around the ordered eigenvalues $\Lambda_0 \leqslant \Lambda_1 \leqslant \Lambda_2 \leqslant \Lambda_3$ of $\hat{L}$. For simplicity of notation, we simply write i for Λ_i. The deeper reason behind this labelling will become clear in the next chapter. The resulting classification is listed in Table 2.2 and compared to the classical results.

Recall that Stäckel systems correspond to projective lines in the KS variety, called Stäckel lines. For each class of separation coordinates the corresponding Stäckel line is given in Table 2.3 via two distinct points on it. Given the trace free symmetric form $\hat{L}$ with eigenvalues $\Lambda_0 \leqslant \Lambda_1 \leqslant \Lambda_2 \leqslant \Lambda_3$, these points can be obtained from (2.38), where n is given by (2.42) as $n_\alpha = \Lambda_0 + \Lambda_\alpha$.

For the next proposition, notice from (2.12) that for an integrable Killing tensor the eigenvalues of the associated algebraic curvature

Table 2.3: Stäckel systems on $\mathbb{S}^3$.

class	two distinct points on the Stäckel line		kernel	characterisation
(0123)	$\begin{pmatrix} 0 & -n_3 & n_2 \\ n_3 & 0 & -n_1 \\ -n_2 & n_1 & 0 \end{pmatrix}$	$\begin{pmatrix} n_2^2 - n_3^2 & -n_1 n_2 & n_3 n_1 \\ n_1 n_2 & n_3^2 - n_1^2 & -n_2 n_3 \\ -n_3 n_1 & n_2 n_3 & n_1^2 - n_2^2 \end{pmatrix}$	$\begin{pmatrix} n_1 \\ n_2 \\ n_3 \end{pmatrix}$	generic case
(01(23))	$\begin{pmatrix} 0 & -n_2 & n_2 \\ n_2 & 0 & -n_1 \\ -n_2 & n_1 & 0 \end{pmatrix}$	$V_{+1} = \begin{pmatrix} 0 & 0 & 0 \\ 0 & +1 & -1 \\ 0 & +1 & -1 \end{pmatrix}$	$\begin{pmatrix} n_1 \\ n_2 \\ n_2 \end{pmatrix}$	$w_2 = w_3$ $t_2 = t_3$
(0(12)3)	$\begin{pmatrix} 0 & -n_3 & n_2 \\ n_3 & 0 & -n_2 \\ -n_2 & n_2 & 0 \end{pmatrix}$	$V_{+3} = \begin{pmatrix} +1 & -1 & 0 \\ +1 & -1 & 0 \\ 0 & 0 & 0 \end{pmatrix}$	$\begin{pmatrix} n_2 \\ n_2 \\ n_3 \end{pmatrix}$	$w_1 = w_2$ $t_1 = t_2$
((01)(23))	$V_{+1} = \begin{pmatrix} 0 & 0 & 0 \\ 0 & +1 & -1 \\ 0 & +1 & -1 \end{pmatrix}$	$V_{-1} = \begin{pmatrix} 0 & 0 & 0 \\ 0 & +1 & +1 \\ 0 & -1 & -1 \end{pmatrix}$	$\begin{pmatrix} 1 \\ 0 \\ 0 \end{pmatrix}$	$w_2 = w_3$ $t_2 = t_3 = 0$
(0(123))	$C_0 = \begin{pmatrix} 0 & -1 & +1 \\ +1 & 0 & -1 \\ -1 & +1 & 0 \end{pmatrix}$	$\begin{pmatrix} n_3 - n_2 & -n_3 & n_2 \\ n_3 & n_1 - n_3 & -n_1 \\ -n_2 & n_1 & n_2 - n_1 \end{pmatrix}$	$\begin{pmatrix} 1 \\ 1 \\ 1 \end{pmatrix}$	$w_\alpha = t_\alpha - \frac{1}{3}(t_1 + t_2 + t_3)$
(0(1(23)))	$C_0 = \begin{pmatrix} 0 & -1 & +1 \\ +1 & 0 & -1 \\ -1 & +1 & 0 \end{pmatrix}$	$V_{+1} = \begin{pmatrix} 0 & 0 & 0 \\ 0 & +1 & -1 \\ 0 & +1 & -1 \end{pmatrix}$	$\begin{pmatrix} 1 \\ 1 \\ 1 \end{pmatrix}$	$w_\alpha = t_\alpha - \frac{1}{3}(t_1 + t_2 + t_3)$ $w_2 = w_3$ $t_2 = t_3$

tensor $\mathbf{R}$, its Weyl part $\mathbf{W}$ and its trace free Ricci part $\mathbf{T}$ are, respectively,

$$\{w_1 \pm t_1, w_2 \pm t_2, w_3 \pm t_3\} \quad \{w_1, w_2, w_3\} \quad \{\pm t_1, \pm t_2, \pm t_3\},$$

where we have omitted the irrelevant scalar part. While the characteristic polynomials of $\mathbf{R}$, $\mathbf{W}$ and $\mathbf{T}$ only yield the respective (unordered) sets of eigenvalues, we can recover the (ordered) triples (w_1, w_2, w_3) and (t_1, t_2, t_3) from these sets, up to a mutual permutation of the indices and up to the signs of the t's.[5] In order to determine the corresponding matrix in the KS variety up to the S_4-symmetry, we still have to determine the sign of the product $t_1 t_2 t_3$. This product is invariantly defined as the Pfaffian of the skew symmetric endomorphism $*\mathbf{T}$, as can be seen from the block-diagonal form (2.3).

Proposition 2.34. *Let $\mathbf{R}$ be the algebraic curvature tensor associated to an integrable Killing tensor on the sphere $\mathbb{S}^3$, $\mathbf{W}$ its Weyl part and $\mathbf{T}$ its trace free Ricci part, seen as endomorphisms on 2-forms. Then the (coefficients of the) characteristic polynomials of $\mathbf{R}$, $\mathbf{W}$ and $\mathbf{T}$ together with the Pfaffian of $*\mathbf{T}$ form a complete set of polynomial invariants on the space of integrable Killing tensors, which are capable of distinguishing the different classes of separation coordinates.*

More precisely, as explained above, we can recover the associated matrix (0.15a) in the KS variety up to the S_4-symmetry. The class of the corresponding separation coordinates is then determined by the following conditions.

(i) $w_\alpha \neq w_\beta = w_\gamma$ and $t_\alpha \neq t_\beta = t_\gamma$ for some permutation (α, β, γ) of $(1, 2, 3)$

(ii) $w_\alpha \neq w_\beta = w_\gamma$ and $t_\alpha \neq t_\beta = t_\gamma = 0$ for some permutation (α, β, γ) of $(1, 2, 3)$

(iii) $w_\alpha = t_\alpha - \frac{1}{3}(t_1 + t_2 + t_3)$ for $\alpha = 1, 2, 3$

---

[5] For a generic integrable Killing tensor it will be sufficient to know the eigenvalues of $\mathbf{W}$ and $\mathbf{T}$ only, since the correctly paired eigenvalues have to satisfy the integrability condition (0.15).

		condition (i)		
		false	true	
condition (ii)	false	elliptic	condition (ii)	
			false	true
			Lamé rotational	cylindrical
	true	Lamé subgroup reduction	spherical	

Proof. This can be directly read off from Table 2.3, in combination with Table 2.2. □

2.6 The space of separation coordinates

From the results in the previous section we can now easily derive the topological description of the space of separation coordinates. In Corollary 2.24 we proved that a Stäckel system on $\mathbb{S}^3$ has mutually diagonalisable algebraic curvature tensors. This justifies the following definition.

Definition 2.35. *We denote by $\mathcal{S}_0(\mathbb{S}^n)$ the variety of Stäckel systems on $\mathbb{S}^n$ with diagonal algebraic curvature tensors.*

Proposition 2.36. *(i) The variety $\mathcal{S}_0(\mathbb{S}^3)$ is isomorphic to the blow up of the projective plane $\mathbb{P}^2$ in the four points $(\pm 1 : \pm 1 : \pm 1)$,*

$$\mathcal{S}_0(\mathbb{S}^3) \cong \mathbb{P}^2 \# 4\mathbb{P}^2. \tag{2.43}$$

(ii) Two Stäckel systems in $\mathcal{S}_0(\mathbb{S}^3)$ are equivalent under isometries if and only if the corresponding points in $\mathbb{P}^2 \# 4\mathbb{P}^2$ are equivalent under the action of the symmetry group S_4 of a tetrahedron in $\mathbb{R}^3$ inscribed in a cube with vertices $(\pm 1, \pm 1, \pm 1)$.

Proof. Recall from Proposition 2.20 that each Stäckel line contains a unique skew symmetric point $\iota(n)$, where $n \in \mathbb{P}^2$, and that each non-singular skew symmetric point $\iota(n)$ determines a unique Stäckel line, given by the two points $\iota(n) \neq \nu(n)$. Moreover, the singular skew symmetric points are the four points $n = (\pm 1 : \pm 1 : \pm 1) \in \mathbb{P}^2$ for which $\iota(n) = \nu(n)$. Hence for the first statement it suffices to show that the subspace spanned by $\iota(n(t))$ and $\nu(n(t))$ has a well defined limit for $t \to 0$ if $n(0)$ is one of these four points and that this limit depends on $\dot{n}(0) \in \mathbb{P}^2$. We leave it to the reader to verify that the limit space is spanned by $\iota(n(0))$ and $\iota(\dot{n}(0))$. The second statement is evident from the S_4-action on the KS variety. $\qquad\square$

Recall that the extension of a Killing tensor by zero defines a map $\mathcal{I}(\mathbb{S}^{n-1}) \hookrightarrow \mathcal{I}(\mathbb{S}^n)$ for every hyperplane $H \subset V$. The extension also preserves commutators, as can be seen, for example, from the algebraic expression (1.40) of the commutator. Hence any Stäckel system on $\mathbb{S}^{n-1}$ extends to an $(n-1)$-dimensional space of commuting integrable Killing tensors on $\mathbb{S}^n$. This space can be completed to a Stäckel system by adjoining the metric on $\mathbb{S}^n$. This defines a map $\mathcal{S}(\mathbb{S}^{n-1}) \hookrightarrow \mathcal{S}(\mathbb{S}^n)$ for each hyperplane $H \subset V$. For $n = 3$ this gives a geometric interpretation of the blow-up.

Proposition 2.37. *The exceptional divisors in the variety (2.43) are Stäckel systems extended from $\mathbb{S}^2 \subset \mathbb{S}^3$ in the sense explained above.*

The non-generic orthogonal separation coordinates are characterised by $|n_1| = |n_2|$, $|n_2| = |n_3|$ and $|n_3| = |n_1|$. This defines six projective lines that divide the projective plane of skew symmetric matrices in the KS variety into twelve congruent triangles. The four singular skew symmetric points are given by $|n_1| = |n_2| = |n_3|$ and each of the triangles has two of them as vertices. Upon blow-up in these points, the triangles become pentagons, which are permuted under the S_4-action. This is shown in Figure 2.2. Since this action is transitive on the set of pentagons and we have twice as many group elements as pentagons, each of the pentagons has a stabiliser isomorphic to $\mathbb{Z}_2$, acting by a reflection. Hence a fundamental domain is given by "half"

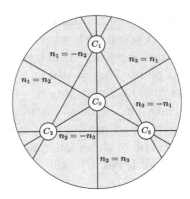

Figure 2.2: $\mathbb{P}^2 \# 4\mathbb{P}^2$ tiled by twelve pentagons.

of the pentagon, which is a quadrilateral. It is not difficult to see that the two vertices of this half not lying on the symmetry axis of the pentagon get identified under the S_4-action.

Proposition 2.38. *The quotient*

$$\mathcal{S}(\mathbb{S}^3)/\operatorname{O}(4) \cong \mathcal{S}_0(\mathbb{S}^3)/S_4$$

is homeomorphic to a quadrilateral with two vertices identified.

We can label the faces of the pentagon by the different classes of separation coordinates derived in the previous section, see Figure 2.3. This is actually the associahedron K_4 and has been the clue for the generalisation in Chapter 3.

2.7 The variety of integrable Killing tensors

In this section we derive the algebraic representation of cofactor systems (see Section 0.7.1) and deduce an algebraic geometric description of the variety $\mathcal{K}(\mathbb{S}^3)$ of integrable Killing tensors on $\mathbb{S}^3$.

The natural way how to define an algebraic curvature tensor out of two symmetric tensors is by by projecting the tensor product

$$\square\square \otimes \square\square \cong \begin{array}{|c|}\hline \\\hline \\\hline\end{array} \oplus \begin{array}{|c|c|}\hline & \\\hline \\\hline\end{array} \oplus \square\square\square\square \longrightarrow \begin{array}{|c|}\hline \\\hline \\\hline\end{array}$$

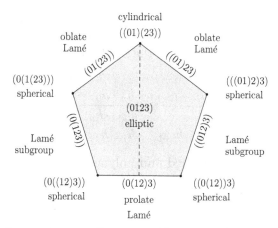

Figure 2.3: Separation coordinates on $\mathbb{S}^3$ parametrised by the associahedron K_4.

onto the component with curvature tensor symmetry. This leads to the following definition.

Definition 2.39. *The* Kulkarni-Nomizu product *of two symmetric tensors h and k on a vector space V is the algebraic curvature tensor $h \oslash k$, given by*

$$(h \oslash k)_{a_1 b_1 a_2 b_2} := h_{a_1 a_2} k_{b_1 b_2} - h_{a_1 b_2} k_{b_1 a_2} - h_{b_1 a_2} k_{a_1 b_2} + h_{b_1 b_2} k_{a_1 a_2}.$$
$$(2.44)$$

Recall that the adjugate matrix $\operatorname{Adj} M$ of a matrix M is the transpose of its cofactor matrix, which is equal to $(\det M)M^{-1}$ if M is invertible. Given a metric, we can identify endomorphisms and bilinear forms. This allows us to extend this definition to bilinear forms. As for endomorphisms, we denote the adjugate of a bilinear form L by $\operatorname{Adj} L$, which is again a bilinear form.

Lemma 2.40. *The algebraic curvature tensor*

$$\frac{(\operatorname{Adj} h) \oslash (\operatorname{Adj} h)}{\det h}, \qquad h \in S^2 V \qquad (2.45)$$

is well defined and polynomial in the components of h.

Proof. Choose a basis in which h is diagonal with diagonal elements $h_i := h_{ii}$. Then $\mathrm{Adj}\, h$ is also diagonal and has diagonal elements $h_0 \cdots h_{i-1} h_{i+1} \cdots h_n$. By (2.44) the algebraic curvature tensor (2.45) is diagonal and given by

$$R_{ijij} = 2\frac{h_0 \cdots h_{i-1} h_{i+1} \cdots h_n \cdot h_0 \cdots h_{j-1} h_{j+1} \cdots h_n}{h_0 \cdots h_n}$$

$$= 2h_0 \cdots h_{i-1} h_{i+1} \cdots h_{j-1} h_{j+1} \cdots h_n$$

for $i \neq j$. This is well defined and polynomial in the components of h. $\qquad\square$

Proposition 2.41 (Algebraic representation of cofactor systems). *Let L be a special conformal Killing tensor on a non-flat constant curvature manifold $M \subset V$, i.e. the restriction of a (constant) symmetric form $\hat{L}$ from V to M. To avoid confusion, we denote the metric on V by $\hat{g}$ and its restriction to $M \subset V$ by g.*

Then the Killing tensors

$$K(\lambda) = \mathrm{Adj}(L - \lambda g), \qquad \lambda \in \mathbb{R} \tag{2.46}$$

in the cofactor system of L correspond to the algebraic curvature tensors

$$R(\lambda) = \frac{\mathrm{Adj}(\hat{L} - \lambda\hat{g}) \oslash \mathrm{Adj}(\hat{L} - \lambda\hat{g})}{\det(\hat{L} - \lambda\hat{g})}. \tag{2.47}$$

Proof. For simplicity of notation, we absorb the trace terms "λg" and "$\lambda\hat{g}$" into L respectively $\hat{L}$ by formally setting λ to 0. At a fixed point $x \in M$ we consider the block decomposition of the endomorphism $\hat{\mathbf{L}}$ under the splitting $V = T_x M \oplus \mathbb{R}x$. Using the blockwise LU decomposition of $\hat{\mathbf{L}}$,

$$\hat{\mathbf{L}} = \begin{pmatrix} \mathbf{L} & l \\ l^t & n \end{pmatrix} = \begin{pmatrix} \mathbf{L} & 0 \\ l^t & 1 \end{pmatrix} \begin{pmatrix} \mathbf{I} & \mathbf{L}^{-1}l \\ 0 & n - l^t\mathbf{L}^{-1}l \end{pmatrix},$$

we compute the block decomposition of the adjugate of $\hat{\mathbf{L}}$ as

$$\mathrm{Adj}\,\hat{\mathbf{L}} = \begin{pmatrix} \dfrac{1}{\det \mathbf{L}}\left(\det \hat{\mathbf{L}}\,\mathrm{Adj}\,\mathbf{L} + (\mathrm{Adj}\,\mathbf{L})ll^t(\mathrm{Adj}\,\mathbf{L})^t\right) & -(\mathrm{Adj}\,\mathbf{L})l \\ -l^t(\mathrm{Adj}\,\mathbf{L})^t & \det \mathbf{L} \end{pmatrix}.$$

Note that we have assumed that $\mathbf{L}$ is invertible, which is possible because (2.46) and (2.47) are polynomial in λ. Now take a basis $e_1, \ldots, e_n$ of $T_x M$ and complete it with $e_0 := x$ to a basis of V. With respect to this basis the components of the Killing tensor K corresponding to the algebraic curvature tensor (2.47) read

$$K_{ij} = R_{0i0j} = \frac{(\operatorname{Adj} \hat{L})_{00} (\operatorname{Adj} \hat{L})_{ij} - (\operatorname{Adj} \hat{L})_{0i} (\operatorname{Adj} \hat{L})_{0j}}{\det \hat{L}} = (\operatorname{Adj} L)_{ij}.$$

That is, K is given by (2.46), as was to be shown. $\qquad \square$

We can now give an algebraic geometric interpretation of the variety of integrable Killing tensors.

Corollary 2.42. *The set of Killing tensors with algebraic curvature tensor of the form $h \otimes h$ for $h \in S^2 V$ and its affine equivalents are dense in $\mathcal{I}(\mathbb{S}^3)$. In other words, the projectivisation $\mathbb{P}\mathcal{I}(\mathbb{S}^3)$ is isomorphic to the projective cone from $g \otimes g$ over the image of the Kulkarni-Nomizu square, a map from the projective space of symmetric forms to the projective space of algebraic curvature tensors, given by $h \mapsto h \otimes h$.*[6] [7]

Proof. Cofactor systems are dense in $\mathcal{S}(\mathbb{S}^3)$, so their union is dense in $\mathcal{I}(\mathbb{S}^3)$. By virtue of the above lemma it suffices therefore to show that the Killing tensors (2.46) and their affine equivalents are dense in the Stäckel system they span. Without loss of generality we can assume the corresponding algebraic curvature tensors (2.47) to be diagonal. Then the composition of the map $\mathbb{P}^1 \to \mathcal{I}(\mathbb{S}^3)$, $\lambda \mapsto K(\lambda)$, with the projection $\mathcal{I}(\mathbb{S}^3) \to \mathcal{I}(\mathbb{S}^3)/\mathrm{Aff}(\mathbb{R})$ defines a regular map $\mathbb{P}^1 \to \mathcal{V}$ from the projective line to the KS variety. The image of this map is contained in a Stäckel line (a projective line) and must therefore be the entire Stäckel line. Otherwise the map would be constant, contradicting that the cofactor system has dimension 3. This proves the first statement of the proposition. For the second statement recall that $g \otimes g$ is the algebraic curvature tensor of the metric (up to a factor). $\qquad \square$

---

[6]This map is 2 : 1 except for rank one symmetric forms h.
[7]Remark that the apex lies itself in this image.

3 The generalisation: a solution for spheres of arbitrary dimension

> In great mathematics there is a very high degree of unexpectedness, combined with inevitability and economy.
>
> GODFREY HAROLD HARDY *(1877 – 1947)*

Contents

3.1 An alternative definition of Stäckel systems

In Definition 0.1, a Killing tensor is a symmetric bilinear form $K_{\alpha\beta}$ on the manifold M. In what follows we will interpret it in two other ways, each of which gives rise to a Lie bracket and hence to a Lie algebra generated by Killing tensors. On one hand, we can use the metric to identify the symmetric bilinear form $K_{\alpha\beta}$ with a symmetric endomorphism $K^{\alpha}{}_{\beta}$. Interpreted in this way, the space of Killing tensors generates a Lie subalgebra of $\Gamma(\mathrm{End}(TM))$ with respect to the commutator bracket (0.4).

On the other hand, we can interpret a Killing tensor $K_{\alpha\beta}$ as a function $K_{\alpha\beta}p^{\alpha}p^{\beta}$ on the total space of the cotangent bundle T^*M which is quadratic in the fibres. Interpreted in this way, the space of Killing tensors generates a Lie subalgebra of $C^{\infty}(T^*M)$ with respect to the Poisson bracket

$$\{K, L\} = \sum_{\alpha=1}^{n} \left(\frac{\partial K}{\partial x^{\alpha}} \frac{\partial L}{\partial p_{\alpha}} - \frac{\partial L}{\partial x^{\alpha}} \frac{\partial K}{\partial p_{\alpha}} \right). \tag{3.1}$$

In this chapter we will use the following alternative definition for Stäckel systems. The equivalence of both definitions is proven in [Ben89].

Definition 3.1. *A Stäckel system on an n-dimensional Riemannian manifold is an n-dimensional space of Killing tensors which mutually commute with respect to both of the following brackets.*

(i) the commutator bracket (0.4)
(ii) the Poisson bracket (3.1)

3.2 Killing tensors with diagonal algebraic curvature tensor

In the following we will only consider Killing tensors on $\mathbb{S}^n$ whose algebraic curvature tensor is diagonal in the following sense of Definition 0.12. Restricting to Killing tensors with diagonal algebraic curvature tensor does not mean any loss of generality. The reason is the following refinement of Theorem 0.5 for spheres.

Theorem 3.2. *[BKW85] Necessary and sufficient conditions for the existence of an orthogonal separable coordinate system for the Hamilton-Jacobi equation on $\mathbb{S}^n$ are that there are n Killing tensors with diagonal algebraic curvature tensor, one of which is the metric, which are linearly independent (locally) and pairwise commute with respect to the Poisson bracket.*

Remark 3.3. *By this theorem condition (ii) in Definition 3.1 implies condition (i) on $\mathbb{S}^n$. As we will show in Section 3.6 below, both conditions are actually equivalent for $\mathbb{S}^n$.*

Theorems 3.2 and 0.10 reduce the classification of separation coordinates on spheres to the purely algebraic problem of finding certain abelian subalgebras in the following two Lie algebras.

Definition 3.4. *We denote by $\mathfrak{d}_{n+1} \subset \Gamma(\mathrm{End}(T\mathbb{S}^n))$ and by $\mathscr{D}_{n+1} \subset C^\infty(T^*\mathbb{S}^n)$ the Lie subalgebras generated by Killing tensors with diagonal algebraic curvature tensor under the commutator bracket (0.4) and the Poisson bracket (3.1) respectively.*

By definition, a diagonal algebraic curvature tensor R is uniquely determined by the diagonal elements R_{ijij} for $1 \leqslant i < j \leqslant n+1$. Indeed, the symmetries (0.6a) and (0.6b) determine the components $R_{ijij} = -R_{ijji} = R_{jiji} = -R_{jiij}$ for $i < j$. And if we set all other components to zero, the resulting tensor R satisfies all symmetries (0.6) of an algebraic curvature tensor. For fixed i and j, let K_{ij} be the Killing tensor with diagonal algebraic curvature tensor R given by

$$R_{ijij} = -R_{ijji} = R_{jiji} = -R_{jiij} = 1 \qquad (3.2)$$

and all other components zero. Then K_{ij} with $i < j$ form a basis of the space of Killing tensors on $\mathbb{S}^n$ with diagonal algebraic curvature tensor and constitute a set of generators for both Lie algebras, $\mathfrak{d}_{n+1}$ and $\mathscr{D}_{n+1}$. The following two propositions show that they satisfy the same relations in $\mathfrak{d}_{n+1}$ and in $\mathscr{D}_{n+1}$.

Proposition 3.5. *Let K_{ij}, $1 \leqslant i < j \leqslant n+1$ be the basis of the space of Killing tensors on $\mathbb{S}^n$ with diagonal algebraic curvature tensor, as defined above. For convenience we set $K_{ji} := K_{ij}$. Then K_{ij} satisfy the following relations in $\mathfrak{O}_{n+1}$.*

$$[K_{ij}, K_{kl}] = 0 \qquad \text{if } i, j, k, l \text{ are distinct} \qquad (3.3a)$$

$$[K_{ij}, K_{ik} + K_{jk}] = 0 \qquad \text{if } i, j, k \text{ are distinct.} \qquad (3.3b)$$

Proof. We can extend the Killing tensor K on $T_x\mathbb{S}^n$ to a symmetric tensor $\hat{K}$ on $V = T_x\mathbb{S}^n \oplus \mathbb{R}x$ by omitting the restriction $v, w \perp x$ in (0.7). The antisymmetry (0.6a) implies that $K_x(v, x) = 0$ for any $v \in V$, so $\hat{K}$ is the extension of K by zero. Consequently, we have $[\hat{K}_{ij}, \hat{K}_{kl}] = \widehat{[K_{ij}, K_{kl}]}$, so that it is sufficient to check the above relations on the corresponding extensions. To do so, consider the Killing tensor K_{ij} at a point $x \in \mathbb{S}^n$. By (0.7) and the definition (3.2) of the diagonal algebraic curvature tensor of K_{ij} we have

$$\hat{K}_{ij}(v, w) = \sum_{a,b=1}^{n+1} \left(R_{abab} x^a x^a v^b w^b + R_{abba} x^a x^b v^b w^a \right)$$

$$= x^i x^i v^j w^j + x^j x^j v^i w^i - x^i x^j v^i w^j - x^i x^j v^j w^i.$$

Let us put all indices down for convenience. Then we have

$$\hat{K}_{ij} = \begin{pmatrix} x_j^2 & -x_i x_j \\ -x_i x_j & x_i^2 \end{pmatrix},$$

where we left only non-zero (i-th and j-th) rows and columns. This already proves relation (3.3a). To check the remaining relation (3.3b), we compute

$$[\hat{K}_{ij}, \hat{K}_{jk}] = \left[\begin{pmatrix} x_j^2 & -x_i x_j & 0 \\ -x_i x_j & x_i^2 & 0 \\ 0 & 0 & 0 \end{pmatrix}, \begin{pmatrix} 0 & 0 & 0 \\ 0 & x_k^2 & -x_j x_k \\ 0 & -x_j x_k & x_j^2 \end{pmatrix} \right]$$

$$= x_i x_j x_k \begin{pmatrix} 0 & -x_k & x_j \\ x_k & 0 & -x_i \\ -x_j & x_i & 0 \end{pmatrix}.$$

$$(3.4)$$

Here we omitted rows and columns other than i, j, k, because they are zero. In the same way we compute $[\hat{K}_{ij}, \hat{K}_{ik}]$ and verify (3.3b). □

Proposition 3.6. *As elements of $\mathscr{D}_{n+1}$ the generators K_{ij} satisfy the following relations.*

$$\{K_{ij}, K_{kl}\} = 0 \qquad \textit{if } i, j, k, l \textit{ are distinct} \qquad (3.5\text{a})$$

$$\{K_{ij}, K_{ik} + K_{jk}\} = 0 \qquad \textit{if } i, j, k \textit{ are distinct.} \qquad (3.5\text{b})$$

Proof. The function on $T^*\mathbb{S}^n$ given by K_{ij} is

$$K_{ij}(x, p) = x_j^2 p_i^2 + x_i^2 p_j^2 - 2x_i x_j p_i p_j = (x_i p_j - x_j p_i)^2. \qquad (3.6\text{a})$$

This already proves relation (3.5a). In order to verify relation (3.5b), we compute

$$\{K_{ij}, K_{jk}\} = \frac{\partial K_{ij}}{\partial x_j} \frac{\partial K_{jk}}{\partial p_j} - \frac{\partial K_{ij}}{\partial p_j} \frac{\partial K_{jk}}{\partial x_j}$$
$$= 4(x_i p_j - x_j p_i)(x_j p_k - x_k p_j)(x_k p_i - x_i p_k), \qquad (3.6\text{b})$$

which is clearly antisymmetric with respect to i and j. □

The next proposition says that there are no more relations between the generators and their brackets, both in $\mathfrak{d}_{n+1}$ as well as in $\mathscr{D}_{n+1}$, provided $n \geqslant 2$.

Proposition 3.7. *The generators K_{ij} and the commutator brackets $[K_{ij}, K_{jk}]$ with $1 \leqslant i < j < k \leqslant n + 1$ are linearly independent in $\mathfrak{d}_{n+1}$ for $n \geqslant 2$. The same is true for K_{ij} and the Poisson brackets $\{K_{ij}, K_{jk}\}$ as elements of $\mathscr{D}_{n+1}$.*

Proof. Since K_{ij} are symmetric and $[K_{ij}, K_{jk}]$ are antisymmetric, it suffices to check the linear independence of both sets independently. The elements K_{ij} are linearly independent by definition. To prove the linear independence of $[K_{ij}, K_{jk}]$ suppose that

$$\sum_{1 \leqslant i < j < k \leqslant n+1} \lambda_{ijk}[K_{ij}, K_{jk}] = 0.$$

For each triple (p, q, r) with $1 \leqslant p < q < r \leqslant n + 1$ consider a point $x \in \mathbb{S}^n$ with $x_m = 0$ if and only if $m \notin \{p, q, r\}$. Due to (3.4) we have that $[K_{ij}, K_{jk}] = 0$ at this point x for all $i < j < k$ unless $(i, j, k) = (p, q, r)$. Hence all $\lambda_{pqr} = 0$.

In the case of $\mathscr{D}_{n+1}$ we note that K_{ij} are quadratic in momenta while $\{K_{ij}, K_{jk}\}$ are cubic; the rest of the proof is the same. $\quad\square$

3.3 The residual action of the isometry group

Theorem 3.2 implies that for a Stäckel system on a sphere we can always find an isometry which takes all Killing tensors in this Stäckel system to Killing tensors having diagonal algebraic curvature tensors. This means that the space of Killing tensors with diagonal algebraic curvature tensor defines a slice for the action of the isometry group. If we want to classify separation coordinates up to isometries, we have to take into account that the stabiliser of this slice in the isometry group is not trivial.

Due to the symmetries (0.6a) and (0.6b), the space of algebraic curvature tensors is a subspace of the space $S^2 \Lambda^2 V$ of symmetric forms on $\Lambda^2 V$. The natural action of the isometry group $\mathrm{O}(V)$ on this space is given as follows. Mapping an orthonormal basis $\{e_i : 1 \leqslant i \leqslant n+1\}$ of V to the basis $\{(e_i \wedge e_j)/\sqrt{2} : 1 \leqslant i < j \leqslant n + 1\}$ of $\Lambda^2 V$ defines a map

$$\mathrm{O}(V) \to \mathrm{O}(\Lambda^2 V), \tag{3.7}$$

since the latter basis is orthonormal with respect to the scalar product on $\Lambda^2 V$ induced from the one on V. Via the action of $\mathrm{O}(\Lambda^2 V)$ on $\Lambda^2 V$ this defines an action of $\mathrm{O}(V)$ on $S^2 \Lambda^2 V$ and hence on algebraic curvature tensors.

In general, the subgroup in $\mathrm{O}(V)$ leaving the space of diagonal bilinear forms on V invariant is the subgroup of signed permutation matrices, acting by permutations and sign changes of the chosen basis $\{e_i\}$ in V. This group is the symmetry group of the hyperoctahedron in V with vertices $\pm e_i$ and is isomorphic to the semidirect product $S_N \ltimes \mathbb{Z}_2^N$, where $N = \dim V = n + 1$.

The stabiliser in $O(V)$ of the space of diagonal algebraic curvature tensors is now the preimage under (3.7) of the stabiliser of diagonal bilinear forms on $\Lambda^2 V$. Since the latter consists of permutations and sign changes of the basis $\{e_i \wedge e_j\}$ in $\Lambda^2 V$, this is just the group of permutations and sign changes of the basis elements e_i, i.e. the group described in the preceding paragraph. Note that the normal subgroup of sign changes, which is isomorphic to $\mathbb{Z}_2^N$, acts trivially on diagonal bilinear forms. Hence the action descends to the quotient $(S_N \ltimes \mathbb{Z}_2^N)/\mathbb{Z}_2^N \cong S_N$. Summarising the above, we have:

Proposition 3.8. *The stabiliser in the isometry group $O(V)$ of the space of Killing tensors on $\mathbb{S}^n \subset V$ with diagonal algebraic curvature tensor is the hyperoctahedral group and isomorphic to the semidirect product $S_N \ltimes \mathbb{Z}_2^N$, where $N = n+1$. This action descends to an action of S_N given by*

$$\sigma(K_{ij}) = K_{\sigma(i)\sigma(j)} \qquad\qquad \sigma \in S_N. \qquad (3.8)$$

3.4 Gaudin subalgebras of the Kohno-Drinfeld Lie algebra and the moduli space $\overline{\mathcal{M}}_{0,n+1}$

We describe now the result of [AFV11], which plays a key role for us.

Definition 3.9. *The (real version of) the Kohno-Drinfeld Lie algebra $\mathfrak{t}_n$ $(n = 2, 3, \ldots)$ is defined as the quotient of the free Lie algebra over $\mathbb{R}$ with generators $t_{ij} = t_{ji}$, $i \neq j \in \{1, \ldots, n\}$, by the ideal generated by the relations*

$$[t_{ij}, t_{kl}] = 0 \qquad \textit{if } i,j,k,l \textit{ are distinct,}$$
$$[t_{ij}, t_{ik} + t_{jk}] = 0 \qquad \textit{if } i,j,k \textit{ are distinct.}$$

This Lie algebra appeared in Kohno's work as the holonomy Lie algebra of the complement to the union of the diagonals $z_i = z_j$, $i < j$ in $\mathbb{C}^n$ (which is also a configuration space of n distinct points on the complex plane) and in Drinfeld's work as the value space of the universal Knizhnik-Zamolodchikov connection (see the references in [AFV11]).

Definition 3.10. A Gaudin subalgebra *of the Kohno-Drinfeld Lie algebra* t_n *is an abelian Lie subalgebra of maximal dimension contained in the linear span* t_n^1 *of the generators* t_{ij}.

Gaudin subalgebras were introduced in [AFV11]. The main class of examples is provided by Gaudin's models of integrable spin chains

$$\mathfrak{g}_n(z) = \left\{ \sum_{1 \leqslant i < j \leqslant n} \frac{b_i - b_j}{z_i - z_j} t_{ij} : \quad b \in \mathbb{R}^n \right\}.$$

Note that they are parametrised by $z \in \Sigma_n / \operatorname{Aff}(\mathbb{R})$, where

$$\Sigma_n = \mathbb{R}^n \setminus \bigcup_{i < j} \{z \in \mathbb{R}^n \mid z_i = z_j\}$$

is the configuration space of n distinct ordered points on the real line and $\operatorname{Aff}(\mathbb{R})$ is the group of affine maps $z \mapsto az + b$, $a \neq 0$, acting diagonally on $\mathbb{R}^n$. A different type of example, which came from the representation theory of the symmetric group, is given by the *Jucys-Murphy subalgebras* spanned by

$$t_{12}, \quad t_{13} + t_{23}, \quad t_{14} + t_{24} + t_{34}, \quad \cdots \tag{3.9}$$

(see [VO04] and references therein).

The main result of [AFV11] is the following.

Theorem 3.11. *[AFV11] Gaudin subalgebras in* t_n *form a nonsingular projective subvariety of the Grassmannian* $G(n-1, n(n-1)/2)$ *of* $(n-1)$-*dimensional subspaces in* t_n^1, *isomorphic to the moduli space* $\overline{\mathcal{M}}_{0,n+1}$ *of stable curves of genus zero with* $n+1$ *marked points.*

In fact, the result holds for any quotient of t_n where both the generators t_{ij}, $1 \leqslant i < j \leqslant n$, and the brackets $[t_{ij}, t_{jk}]$, $1 \leqslant i < j < k \leqslant n$, are linearly independent (see remark 2.6 in [AFV11]), and over any field.

The most popular version of the moduli space $\overline{\mathcal{M}}_{0,n+1}$ – appearing, for example, in the celebrated Witten's conjecture – is defined over $\mathbb{C}$. It is a particular (Deligne-Mumford) compactification of the configuration space $\bar{\mathcal{M}}_{0,n+1}(\mathbb{C})$ of $n+1$ distinct labelled points in $\mathbb{C}P^1$ modulo $PGL_2(\mathbb{C})$ studied by Knudsen [Knu83], who proved that it is a smooth projective variety. The compactification $\bar{\mathcal{M}}_{0,n+1}(\mathbb{C})$ includes the singular rational curves with double point singularities and with the following properties: the graph of components is a tree (genus zero) and each irreducible component contains at least three marked or singular points (stability condition).

However, we need the real version $\bar{\mathcal{M}}_{0,n+1}(\mathbb{R})$, which we discuss next.

3.5 The real version $\bar{\mathcal{M}}_{0,n+1}(\mathbb{R})$ and Stasheff polytopes

3.5.1 Topology

The real version $\bar{\mathcal{M}}_{0,n+1}(\mathbb{R})$ was studied in more detail by Kapranov [Kap93] and Devadoss [Dev99]. By Knudsen's theorem, which works over $\mathbb{R}$ as well, $\bar{\mathcal{M}}_{0,n+1}(\mathbb{R})$ is a smooth real manifold of dimension $n-2$. It can be described as an iterated blow-up of $\mathbb{R}P^{n-2}$ [Kap93, Dev99, DJS98]. $\bar{\mathcal{M}}_{0,4}(\mathbb{R})$ is simply $\mathbb{R}P^1$ and $\bar{\mathcal{M}}_{0,5}(\mathbb{R})$ is a non-orientable surface with Euler characteristic -3, which is a connected sum of five copies of $\mathbb{R}P^2$.

The topology of $\bar{\mathcal{M}}_{0,n+1}(\mathbb{R})$ becomes increasingly complicated when n grows. It is known to be aspherical (Davis et al. [DJS98]). The Euler characteristic can be given explicitly by

$$\chi\big(\bar{\mathcal{M}}_{0,n+1}(\mathbb{R})\big) = (-1)^{\frac{n-2}{2}}(n-1)!!(n-3)!!$$

for even n (and zero for odd n), see [Dev99]. A description of the cohomology is more complicated than in the complex case, as found by Etingof et al. in [EHKR10].

3.5.2 Combinatorics

Fortunately, a lot of information about $\bar{\mathcal{M}}_{0,n+1}(\mathbb{R})$ is encapsulated in a well studied remarkable polytope known as *associahedron*, or

Stasheff polytope K_n. Namely, $\bar{\mathcal{M}}_{0,n+1}(\mathbb{R})$ is tessellated by $n!/2$ copies of K_n, see [Kap93, Dev99].

K_n was first described by Stasheff as a combinatorial object in the homotopy theory of H-spaces [Sta63] (see the history of this in Stasheff's contribution to [MHPS12]). Its first realisation as a convex polytope is usually ascribed to Milnor. By now we have several geometric realisations of Stasheff polytopes, see e.g. [Dev09] and references therein. The one used for Figure 3.1 is from [Lod04]. K_n is a convex polytope of dimension $n-2$: K_3 is a segment, K_4 is a pentagon and K_5 is the polyhedron shown in Figure 3.1, which can be obtained combinatorially by cutting off three skew edges from a cube.

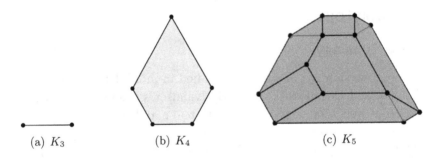

(a) K_3 (b) K_4 (c) K_5

Figure 3.1: Stasheff polytopes.

The faces of K_n of codimension d are in one-to-one correspondence with dissections of a based $(n+1)$-gon by d non-intersecting diagonals (see e. g. [DR01]). In particular, the vertices of K_n correspond to the triangulations of the $(n+1)$-gon by non-intersecting diagonals and their number is C_{n-1}, where

$$C_n = \frac{1}{n+1}\binom{2n}{n}$$

is the *Catalan number*.

Alternatively, the faces of K_n can be labelled by non-isomorphic planar rooted trees with n leaves (see e. g. [Dev99]). These are simply the dual graphs of the dissected polygons, cut off at the edges of the polygon. In particular, the vertices of K_n correspond to binary rooted trees. For $n = 4$ this is depicted in Figure 3.2.

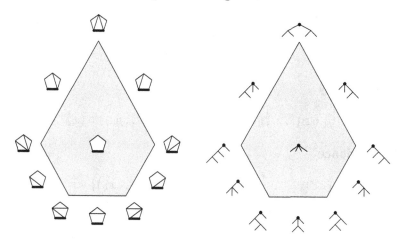

Figure 3.2: Labellings of K_4 by dissections of a based pentagon (left) and planar rooted trees with four leaves (right).

The Stasheff polytope K_n admits a realisation with the dihedral symmetry D_{n+1}, which is the symmetry group of a regular $(n+1)$-gon [Lee89].

3.5.3 Operad structure

The sequence of moduli spaces $\bar{\mathcal{M}}_{0,n+1}(\mathbb{R})$ for $n = 1, 2, \ldots$ carries a natural operad structure, called the "mosaic operad" [Dev99].

Definition 3.12. *An* operad structure *on a sequence of objects* $\mathcal{O}(n)$ *is a composition map*

$$\circ: \quad \mathcal{O}(k) \times \mathcal{O}(n_1) \times \cdots \times \mathcal{O}(n_k) \quad \longrightarrow \quad \mathcal{O}(n_1 + \cdots + n_k)$$
$$(y, x_1, \ldots, x_k) \quad \longmapsto \quad y \circ (x_1, \ldots, x_k)$$

together with a right action

$$*: \quad \mathcal{O}(n) \times S_n \quad \longrightarrow \quad \mathcal{O}(n)$$
$$(x, \pi) \quad \longmapsto \quad x * \pi$$

of the permutation group S_n on each object $\mathcal{O}(n)$, satisfying the following axioms.

Identity There is a distinguished element $1 \in \mathcal{O}(1)$ satisfying

$$y \circ (1, \dots, 1) = y = 1 \circ y.$$

Associativity

$$z \circ (y_1 \circ x_1, \dots, y_k \circ x_k) = \big(z \circ (y_1, \dots, y_k)\big) \circ (x_1, \dots, x_k),$$

Equivariance

$$(y * \pi) \circ ((x_1, \dots, x_k) * \pi) = \big(y \circ (x_1, \dots, x_k)\big) * \pi$$
$$y \circ (x_1 * \pi_1, \dots, x_k * \pi_k) = \big(y \circ (x_1, \dots, x_k)\big) * (\pi_1, \dots, \pi_k),$$

where S_k acts on $(x_1, \dots, x_k)$ by permutation and $(\pi_1, \dots, \pi_k)$ on $\mathcal{O}(n_1 + \dots + n_k)$ under the inclusion $S_{n_1} \times \dots \times S_{n_k} \hookrightarrow S_{n_1 + \dots + n_k}$.

In terms of dissected polygons, the operad structure on $\mathcal{O}(n) := \bar{\mathcal{M}}_{0,n+1}(\mathbb{R})$ is given by gluing the k $(n_i + 1)$-gons x_i with their base to the k non-base edges of the $(k+1)$-gon y to form the $(n_1 + \dots + n_k + 1)$-gon $y \circ (x_1, \dots, x_k)$ with the base of y as base. If y is dissected by d_0 diagonals and x_i by d_i diagonals, then $y \circ (x_1, \dots, x_k)$ is dissected by $d_0 + d_1 + \dots d_k + k$ diagonals, namely the diagonals of y and $x_1, \dots, x_k$ plus the k glued pairs of edges which become diagonals after gluing. On planar trees, the composition $y \circ (x_1, \dots, x_k)$ is given by grafting the k trees x_i with their root to the leaves of the tree y.

The operad structure on Stasheff polytopes defines a map

$$\circ : K_k \times K_{n_1} \times \dots \times K_{n_k} \hookrightarrow K_{n_1 + \dots + n_k}$$

whose image is a codimension k face of $K_{n_1 + \dots + n_k}$. This yields a decomposition of the faces of a Stasheff polytope into products of Stasheff polytopes [Sta63].

3.6 The correspondence

After these preparations we are now in a position to state our main result. By Propositions 3.5 and 3.6 the defining relations of the Kohno-Drinfeld Lie algebra $\mathfrak{t}_n$ are satisfied in the Lie algebras $\mathfrak{d}_n$ and $\mathscr{D}_n$ (cf. Definition 3.4). This provides surjective Lie algebra morphisms

$$\mathfrak{t}_n \longrightarrow \mathfrak{d}_n \qquad\qquad \mathfrak{t}_n \longrightarrow \mathscr{D}_n, \qquad (3.10)$$

given by mapping the generator t_{ij} to the Killing tensor K_{ij}. Under these morphisms the linear span $\mathfrak{t}_n^1$ of the t_{ij} is isomorphic to the space $\mathcal{K}(\mathbb{S}^{n-1})$ of Killing tensors on $\mathbb{S}^{n-1}$ – interpreted as endomorphisms in $\mathfrak{d}_n$ respectively as quadratic functions on the cotangent bundle in $\mathscr{D}_n$. Thus the above morphisms map Gaudin subalgebras in the Kohno-Drinfeld Lie algebra $\mathfrak{t}_n$ to Stäckel systems on $\mathbb{S}^{n-1}$ with diagonal algebraic curvature tensor. Proposition 3.7 now shows that this defines an isomorphism between Gaudin subalgebras and Stäckel systems. Now using Theorem 3.11 we have the following correspondence.

Theorem 3.13. *The Stäckel systems on $\mathbb{S}^n$ with diagonal algebraic curvature tensor form a nonsingular algebraic subvariety of the Grassmannian $G(n, n(n+1)/2)$ of n-planes in the space of Killing tensors with diagonal algebraic curvature tensor, which is isomorphic to the real Deligne-Mumford-Knudsen moduli space $\bar{\mathcal{M}}_{0,n+2}(\mathbb{R})$ of stable genus zero curves with $n+2$ marked points.*

Note that since $\bar{\mathcal{M}}_{0,n+2}(\mathbb{R})$ can be considered as a compactification of the configuration space $\Sigma_{n+1}/\operatorname{Aff}(\mathbb{R})$ of $n+1$ ordered distinct points on a real line modulo the affine group, we have a natural action of the symmetric group S_{n+1} on $\bar{\mathcal{M}}_{0,n+2}(\mathbb{R})$.

Corollary 3.14. *The space $\mathcal{S}(\mathbb{S}^n)/\operatorname{O}(n+1)$ of equivalence classes of orthogonal separation coordinates on the sphere $\mathbb{S}^n$ modulo the isometry group and the quotient space $\bar{\mathcal{M}}_{0,n+2}(\mathbb{R})/S_{n+1}$ are naturally homeomorphic.*

Proof. This follows directly from Theorems 0.5 and 3.13. It suffices to note that by Proposition 3.8 the morphisms (3.10) are equivariant with respect to the S_{n+1}-action on $\mathfrak{t}_{n+1}$ and $\mathfrak{d}_{n+1}$ respectively $\mathscr{D}_{n+1}$. Therefore the isomorphism in Theorem 3.13 is S_{n+1}-equivariant, so that the corresponding quotients are homeomorphic. Note that while the space of Stäckel systems on $\mathbb{S}^n \subset V$ with diagonal algebraic curvature tensor depends on the choice of an orthonormal basis in the ambient space V (for which the algebraic curvature tensors are diagonal), the quotient does not. Therefore the homeomorphism is natural. $\qquad\square$

Since $\bar{\mathscr{M}}_{0,n+2}(\mathbb{R})$ is tessellated by $(n+1)!/2$ copies of the Stasheff polytope K_{n+1}, we can use it to describe the quotient. The interior of K_{n+1} corresponds to the classical elliptic coordinates (0.11) on the sphere $\mathbb{S}^n$. The $n+1$ distinct real parameters $(\Lambda_1, \ldots, \Lambda_{n+1}) \in \Sigma_{n+1}$ they depend on are the eigenvalues of the symmetric form $\hat{L}$ on $\mathbb{R}^{n+1}$ which restricts to the corresponding special conformal Killing tensor L on $\mathbb{S}^n$. Shifting or scaling them only reparametrises the corresponding coordinates. Hence the actual parameter space is the quotient $\Sigma_{n+1}/\operatorname{Aff}(\mathbb{R})$, which is nothing else but the open moduli space $\mathscr{M}_{0,n+2}(\mathbb{R})$. Thus we have the following important corollary.

Die Grenze ist der eigentlich fruchtbare Ort der Erkenntnis.

PAUL TILLICH *(1886 – 1965)*
theologician and philosopher

Corollary 3.15. *All orthogonal separation coordinates on $\mathbb{S}^n$ belong to the closure of the Neumann family of elliptic coordinates. The possible degenerations of the Neumann family correspond to the faces of the Stasheff polytope K_{n+1}.*

The first part is probably not surprising for the experts (see the similar claim in the complex case in [KMR84]), but we have not seen it explicitly stated and proved in the literature. In Section 3.7.3 we will show that rather than by actually performing the limiting process explicitly (as in [KMR84]), the limiting cases can be better understood by composing generic separation coordinates (that is elliptic coordinates) of lower dimensions under the operad composition. The same holds true for the corresponding Stäckel systems.

Because we have $(n + 1)!/2$ Stasheff polytopes K_{n+1} tiling the moduli space $\bar{\mathcal{M}}_{0,n+2}(\mathbb{R})$, the quotient $\bar{\mathcal{M}}_{0,n+2}(\mathbb{R})/S_{n+1}$ is actually only "a half" of K_{n+1} with some identification between the faces. In the interior of the polytope the identification is given by the action of $\mathbb{Z}_2 \subset D_{n+2}$, corresponding to a reflection in the dihedral group, realised as an isometry of K_{n+1} (see above). If we are using the blow-up description of $\bar{\mathcal{M}}_{0,n+2}(\mathbb{R})$ [Kap93, Dev99, DJS98], then it corresponds to the longest element $(1, 2, \ldots, n, n + 1) \mapsto (n + 1, n, \ldots, 2, 1)$ in the symmetric group S_{n+1}, mapping the A_n Weyl chamber into the opposite one. For K_4 this is just a reflection symmetry of the pentagon as indicated in Figure 3.4.

The identification of the faces is more sophisticated. Probably the best way to describe them is using the "twisting along the diagonal" operation for the dissected $(n + 1)$-gon introduced in [Dev99, DR01]. On planar trees this corresponds to reversing the ordering of a node's children. Trees which are equivalent under this operation are called "dyslexic".

As a matter of illustration, let us consider the least non-trivial example $n = 2$, depicted in Figure 3.3. The parameter space $\bar{\mathcal{M}}_{0,4}(\mathbb{R}) \cong \mathbb{RP}^1$ is just a circle, parametrised by the affine-invariant cross ratio $\tau = \frac{\Lambda_2 - \Lambda_1}{\Lambda_3 - \Lambda_1}$. It is tessellated by three copies of the Stasheff polytope K_3, the tiles being the arcs between the points $\tau = 0, 1, \infty$. We can identify K_3 with the arc $[0, 1]$, for which $\Lambda_1 < \Lambda_2 < \Lambda_3$. The symmetry group S_3 acts by permuting Λ_1, Λ_2 and Λ_3 and hence the points 0, 1 and ∞. The quotient $\bar{\mathcal{M}}_{0,4}(\mathbb{R})/S_3$ can be identified with the arc $[0, 1/2]$. This is "a half" of K_3, where $\tau = 0$ corresponds to spherical coordinates and $\tau = 1/2$ to the "lemniscatic" case, which is just a particular case of elliptic coordinates when $\Lambda_2 = \frac{\Lambda_1 + \Lambda_3}{2}$. In this example there is no identification in place, because the dimension is too low.

Note that there is a $\mathbb{Z}_2$ monodromy present in this moduli space, because we considered separation coordinates as an *unordered* system

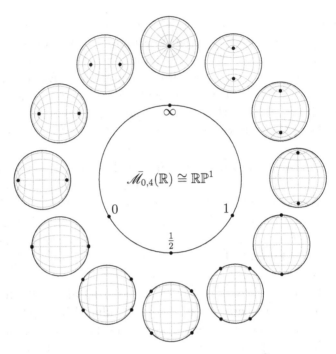

Figure 3.3: Orthogonal separation coordinates on $\mathbb{S}^2$ parametrised by $\bar{\mathscr{M}}_{0,4}(\mathbb{R}) \cong \mathbb{RP}^1$.

of coordinate hypersurfaces: Starting from spherical coordinates and going round once in $\bar{\mathscr{M}}_{0,4}(\mathbb{R}) \cong \mathbb{RP}^1$, longitude and latitude exchange. This is illustrated on the cover page of this book by colouring the individual coordinates and using the double cover $\mathbb{S}^1 \to \mathbb{RP}^1$ for a parametrisation.

3.7 Applications

3.7.1 Enumerating separation coordinates

For a Stasheff polytope, the number of non-equivalent faces of a given dimension can be given by the following Devadoss-Read formula

[DR01]. Let $A(x, y) = \sum a_{mn} x^m y^n$ be the formal series solution of the functional equation

$$A(x, y) = y + \frac{1}{2} \left(\frac{A(x, y)^2}{1 - A(x, y)} + \frac{(1 + A(x, y))A(x^2, y^2)}{1 - A(x^2, y^2)} \right). \quad (3.11)$$

There is no closed formula for $A(x, y)$, but one uses Equation (3.11) to find the coefficients a_{mn} recursively. The claim is that the coefficient a_{mn} is the number of non-equivalent faces of K_n of codimension $m - 1$. Devadoss and Read proved this using a combinatorial technique going back to Pólya [DR01].

Using Table 2 from [DR01], we get the number of non-equivalent canonical forms of separation coordinates on $\mathbb{S}^n$ for $n \leqslant 10$, as listed in Table 3.1. The 1's on the diagonal correspond to elliptic coordinates,

Table 3.1: Number of canonical forms for separation coordinates on $\mathbb{S}^n$, ordered by increasing number of independent continuous parameters.

	0	1	2	3	4	5	6	7	8	9	total
$\mathbb{S}^2$	1	1									2
$\mathbb{S}^3$	2	3	1								6
$\mathbb{S}^4$	3	8	5	1							17
$\mathbb{S}^5$	6	20	22	8	1						57
$\mathbb{S}^6$	11	49	73	46	11	1					191
$\mathbb{S}^7$	23	119	233	206	87	15	1				684
$\mathbb{S}^8$	46	288	689	807	485	147	19	1			2482
$\mathbb{S}^9$	98	696	1988	2891	2320	1021	236	24	1		9275
$\mathbb{S}^{10}$	207	1681	5561	9737	9800	5795	1960	356	29	1	35127

the numbers in the first column to polyspherical coordinates. Note that the sequence $1, 2, 3, 6, 11, 23, \ldots$ is the sequence of Wedderburn-Etherington numbers, i.e. the number of non-planar binary rooted trees with $n + 1$ leaves [OEISa]. This reflects the fact that we can parametrise polyspherical coordinates by planar binary rooted trees and that the notions *dyslexic* and *non-planar* coincide for binary trees.

In the first row we have 1 and 1, corresponding to spherical and elliptic coordinates on $\mathbb{S}^2$ respectively, as discussed in Section 3.6. The numbers in the second row – 2, 3, 1 – are in perfect agreement with the classical results of Eisenhart [Eis34], Olevskiĭ[Ole50] and Kalnins & Miller [KM86]. They correspond to spherical and cylindrical coordinates (2), two types of Lamé rotational coordinates plus Lamé subgroup reduction (3) and elliptic coordinates (1) respectively. Their identification with the faces of the Stasheff polytope K_4 is indicated in Figure 3.4, in comparison to the different labellings shown in Figure 3.2. Polyspherical coordinates comprise, for example, the usual spherical coordinates plus cylindrical coordinates. Note that "the

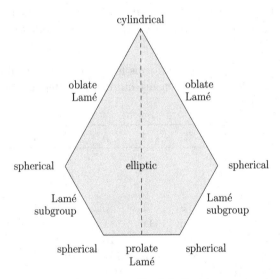

Figure 3.4: The Stasheff polytope K_4, labelled by separation coordinates on $\mathbb{S}^3$.

half" of K_4, obtained as the quotient under the reflection symmetry indicated in Figure 3.4, is a quadrilateral and that for the quotient space $\bar{\mathscr{M}}_{0,5}(\mathbb{R})/S_4$ we have to identify its two adjacent vertices that are labelled by spherical coordinates and joined by Lamé subgroup reduction.

In accordance with our description, the total number of canonical forms for separation coordinates on $\mathbb{S}^n$, indicated in the last column in Table 3.1, is the number of dyslexic planar rooted trees with $n + 1$ leaves [OEISb].

3.7.2 Constructing separation coordinates via the mosaic operad

The correspondence in Theorem 3.13 transfers the natural operad structure on $\overline{\mathcal{M}}_{0,n+2}(\mathbb{R})$ to orthogonal separation coordinates on $\mathbb{S}^n$ and thereby yields a uniform construction procedure for separation coordinates on spheres. Implicitly this structure is already present in [KM86] (see, in particular, formula (3.14) therein). Let us now make this operad structure explicit. We begin with the operad structure on $\mathcal{O}(n) := \mathbb{R}^n$ given by

$$\circ: \quad \mathbb{R}^k \times \mathbb{R}^{n_1} \times \ldots \times \mathbb{R}^{n_k} \longrightarrow \mathbb{R}^{n_1 + \cdots + n_k}$$
$$(\boldsymbol{y}, \boldsymbol{x}_1, \ldots, \boldsymbol{x}_k) \mapsto \boldsymbol{y} \circ (\boldsymbol{x}_1, \ldots, \boldsymbol{x}_k) := (y_1 \boldsymbol{x}_1, \ldots, y_k \boldsymbol{x}_k)$$

together with the permutation action of S_n on $\mathbb{R}^n$. This operad structure descends to an operad structure on $\mathcal{O}(n) := \mathbb{S}^{n-1} \subset \mathbb{R}^n$, since $(y_1 \boldsymbol{x}_1)^2 + \cdots (y_k \boldsymbol{x}_k)^2 = y_1^2 + \cdots y_k^2 = 1$ for $\boldsymbol{y} \in S^{k-1}$ and $\boldsymbol{x}_\alpha \in \mathbb{S}^{n_\alpha - 1}$. Note that the composition map

$$\circ: \mathbb{S}^{k-1} \times \mathbb{S}^{n_1 - 1} \times \cdots \times \mathbb{S}^{n_k - 1} \longrightarrow \mathbb{S}^{n_1 + \cdots + n_k - 1}$$

describes the k-fold join $\mathbb{S}^{n_1 - 1} \star \cdots \star \mathbb{S}^{n_k - 1} \cong \mathbb{S}^{n_1 + \cdots + n_k - 1}$ of the spheres $\mathbb{S}^{n_1 - 1}, \ldots, \mathbb{S}^{n_k - 1}$. This operad structure on spheres induces an operad structure on (local) coordinates on spheres, since coordinates

$$\boldsymbol{x}_0 = \boldsymbol{x}_0(\varphi_{01}, \ldots, \varphi_{0,k-1}) \quad \text{on} \quad \mathbb{S}^{k-1} \subset \mathbb{R}^k \quad \text{and}$$
$$\boldsymbol{x}_\alpha = \boldsymbol{x}_\alpha(\varphi_{\alpha 1}, \ldots, \varphi_{\alpha, n_\alpha - 1}) \quad \text{on} \quad \mathbb{S}^{n_\alpha - 1} \subset \mathbb{R}^{n_\alpha} \quad \alpha = 1, \ldots, k$$

determine coordinates

$$\boldsymbol{x} = \boldsymbol{x}(\varphi_{01}, \ldots, \varphi_{0,k-1}, \varphi_{11}, \ldots, \varphi_{1,n_1 - 1}, \ldots, \varphi_{k1}, \ldots, \varphi_{k,n_k - 1})$$

on $\mathbb{S}^{n_1+\cdots+n_k-1}$, given by setting

$$x := x_0 \circ (x_1, \ldots, x_k). \tag{3.12}$$

The interior of a Stasheff polytope corresponds to elliptic coordinates and its faces are products of Stasheff polytopes. Therefore we can construct all orthogonal separation coordinates on spheres (modulo isometries) from elliptic coordinates by composing them in a recursive manner via the operad composition (3.12). Just start with trivial coordinates $x(\varnothing) = 1$ on a certain number of zero dimensional spheres $\mathbb{S}^0$ and take elliptic coordinates for x_0 in each step. The different choices one has when iterating this composition are given by the trees labelling the corresponding separation coordinates. That is, the rooted trees describe the hierarchy of iterated decompositions of a sphere as joins of lower dimensional spheres. This parallels the decomposition of the faces of a Stasheff polytope into products of lower dimensional Stasheff polytopes.

Note that the construction (0.12) of Vilenkin's polyspherical coordinates on $\mathbb{S}^{n-1}$ corresponds to the special case $k = 2$ of the above construction, starting from trivial coordinates $x(\varnothing) = 1$ on n copies of $\mathbb{S}^0$ and using the (elliptic) coordinates $x_0(\varphi) = (\cos\varphi, \sin\varphi)$ on $\mathbb{S}^{k-1} = \mathbb{S}^1$ in each step.

Moreover, this operad structure on separation coordinates also explains Kalnins & Miller's graphical procedure [KM86]. Namely, adding in an "irreducible block" a leaf to each box which is not joined to another block and replacing each irreducible block by a node, the graphs in [KM86] become the trees arising from the operad structure.

3.7.3 Constructing Stäckel systems via the mosaic operad

We now explain how this operad structure manifests itself on the level of Stäckel systems. To this end, let $I_1 \cup \cdots \cup I_k = I$ be a partition of $I = \{1, \ldots, n\}$ with $|I_\alpha| =: n_\alpha$ and set $I_0 := \{1, \ldots, k\}$. We denote by $\mathfrak{d}_n^1$ the space of Killing tensors on $\mathbb{S}^{n-1}$ with diagonal algebraic curvature tensor and define the injections

$$\iota_0: \quad \mathfrak{d}_k^1 \hookrightarrow \mathfrak{d}_{n_1+\cdots+n_k}^1 \qquad \iota_0(K_{\alpha\beta}) := \sum_{a \in I_\alpha, b \in I_\beta} K_{ab} \qquad \alpha, \beta \in I_0$$

$$\iota_\alpha: \quad \mathfrak{d}_{n_\alpha}^1 \hookrightarrow \mathfrak{d}_{n_1+\cdots+n_k}^1 \qquad \iota_\alpha(K_{ij}) := K_{ij} \qquad i, j \in I_\alpha.$$

Proposition 3.16. *Let Σ_0 be a Stäckel system on $\mathbb{S}^{k-1}$ and Σ_α be Stäckel systems on $\mathbb{S}^{n_\alpha-1}$ for $\alpha = 1, \ldots, k$, all consisting of Killing tensors with diagonal algebraic curvature tensor. Then*

$$\Sigma_0 \circ (\Sigma_1, \ldots, \Sigma_k) := \iota_0(\Sigma_0) \oplus \iota_1(\Sigma_1) \oplus \cdots \oplus \iota_k(\Sigma_k) \qquad (3.13)$$

is a Stäckel system on $\mathbb{S}^{n_1+\cdots+n_k-1}$. Moreover, this operation together with the S_n-action (3.8) defines an operad structure on those Stäckel systems on $\mathbb{S}^{n-1}$ that consist of Killing tensors with diagonal algebraic curvature tensor.

Proof. First observe that the sum on the right hand side of (3.13) is indeed a direct sum. Hence its dimension is $n_1 + \cdots + n_k - 1$. By Definition 3.1 and Remark 3.3, we have to show that all Killing tensors in this subspace commute. That is, a Killing tensor from $\iota_p(\Sigma_p)$ and another one from $\iota_q(\Sigma_q)$ commute for all $p, q = 0, 1, \ldots, k$. For $p, q \neq 0$ this is evident. For $p = q = 0$ one readily checks that the inclusion ι_0 preserves the relations (3.3) and hence maps commuting Killing tensors to commuting Killing tensors. In the remaining case $p \neq 0 = q$ the commutator

$$[\iota_p(K_{ij}), \iota_0(K_{\alpha\beta})] = [K_{ij}, \sum_{a \in I_\alpha, b \in I_\beta} K_{ab}] \qquad i, j \in I_p \qquad \alpha \neq \beta \in I_0$$

is zero unless $p = \alpha$ or $p = \beta$. But if $p = \alpha$, the sum over $a \in I_\alpha$ only contributes non-zero terms for $a = i$ and $a = j$. So the above commutator reduces to

$$\sum_{b \in I_\alpha} [K_{ij}, K_{ib} + K_{jb}] = 0$$

due to the relations (3.3), and similarly for $p = \beta$. This proves that (3.13) is a Stäckel system.

To check that this composition defines an operad is straightforward. The identity element is the empty Stäckel system on $\mathbb{S}^0$ and equivariance is obvious. Associativity can be shown by taking subdivisions $I_\alpha = I_{\alpha 1} \cup \cdots \cup I_{\alpha k_\alpha}$ of I_α for all $\alpha \in I_0$ and considering the corresponding inclusions for Killing tensors. The details will be left to the reader. $\square$

To give an example, let us construct the Stäckel system for standard spherical coordinates on $\mathbb{S}^{n-1}$ by choosing $k = 2$ with the Stäckel system Σ_0 on $\mathbb{S}^{k-1} = \mathbb{S}^1$ spanned by K_{12}, starting from empty Stäckel systems on n copies of $\mathbb{S}^0$ and taking $n_2 = 1$ in each step. This yields the Stäckel system spanned by

$$K_{12}, \quad K_{13} + K_{23}, \quad K_{14} + K_{24} + K_{34}, \quad \cdots$$

and shows that the Jucys-Murphy subalgebras (3.9) in the Kohno-Drinfeld Lie algebra correspond to standard spherical coordinates.

4 The perspectives: applications and generalisations

Never measure the height of a mountain
until you have reached the top.
Then you will see how low it was.

DAG HJALMAR AGNE CARL HAMMARSKJÖLD *(1905
– 1961)*
Nobel Peace Prize laureate 1961

Contents

We have proposed a new, purely algebraic geometric approach to the problem of separation of variables and we have demonstrated that this approach is viable by successfully carrying it out for the simplest non-trivial family of examples – that of spheres. In particular, we elucidated the natural algebro-geometric structure of the parameter space classifying equivalence classes of separation coordinates, which for a long time had only been known as a mere set, and gave a precise description of its topology. In this way we discovered that the theory of Deligne-Mumford-Knudsen moduli spaces and Stasheff polytopes provides the right framework for the classification and construction of all orthogonal separation coordinates on spheres.

Our achievement can only be a starting point. On one hand, spheres are the simplest family of Riemannian manifolds admitting non-trivial moduli spaces of separation coordinates. Applying our approach to other families of Riemannian manifolds and to other notions of separation of variables will yield many new examples of interesting moduli spaces. The hope is to be able to identify them to moduli spaces known from other fields of mathematics and to exploit such a correspondence in one direction or the other. On the other hand, a complete classification of separation coordinates is just the first step towards an explicit solution of the partial differential equation in question or towards a description of the dynamics of the integrable systems described by the corresponding Stäckel systems. We propose to reconsider classical problems related to separation of variables from a consequent algebraic geometric point of view, based on our results.

In what remains, we would like to illustrate this with a brief explanation of some of the possible extensions and applications of our approach, accompanied by a number of problems which arise naturally from this context.

4.1 Other families of Riemannian manifolds

Although only carried out for spheres, our approach is generic for all other constant curvature manifolds, even in the pseudo-Riemannian setting. Due to their non-compact isometry groups, these cases are more involved. This difference only imposes technical problems, not fundamental ones.

Hyperbolic space.   For hyperbolic space $\mathbb{H}^n \subset V$, the only difference in the algebraic integrability conditions (0.13) is that the ambient space V is Lorentzian. The main difficulties in this case arise from the fact that symmetric forms in Lorentzian signature are not diagonalisable under the pseudo-orthogonal group. Instead, we have four different classes of normal forms for the Ricci tensor, corresponding to the four classes of separation coordinates named A, B, C and D in Kalnin's classification [Kal86]. Accordingly, there will be four (intersecting) slices and a different algebraic variety in each of them.

Class A contains diagonalisable Ricci tensors and is thus completely analogous to the case of spheres. Stäckel systems of class A on $\mathbb{H}^n$ will also be parametrised by the moduli spaces $\bar{\mathcal{M}}_{0,n+2}(\mathbb{R})$ and we can use Stasheff polytopes to classify separation coordinates of class A. The essential difference is that the stabiliser of slice A is not S_{n+1}, as it has to preserve the timelike direction of the Lorentz space V.

Class B contains Ricci tensors which are diagonalisable under the complex orthogonal group, so it is almost analogous to the case of spheres. The KS variety in slice B is a linear determinantal variety determined by the same equation as in the real case, with the difference that instead of real variables we have complex conjugated pairs. Hence class B will be parametrised by a different real form of the moduli space $\bar{\mathcal{M}}_{0,n+2}(\mathbb{C})$, determined by a different real structure (see below). It remains to be investigated whether this real form also carries some tiling by polytopes or an operad structure.

The "slices" for classes C and D are not slices in the proper sense, as here the remainder of the isometry group action is not discrete. The parameter space of separation coordinates on $\mathbb{H}^n$ will not even be Hausdorff. We should point out that spheres are really exceptional (and probably unique) in that their separation coordinates are parametrised by a *smooth* projective variety and that the quotient under isometries has a simple orbifold structure. As known from other moduli problems, moduli spaces are in general much more complicated objects: they will contain several components, singularities or even non-separable points. Here we see that already for hyperbolic space all these phenomena will actually occur.

Euclidean space. The situation is similar for Euclidean space, which is the most interesting for applications in physics. Here diagonalisability also fails, as we have to deal with normal forms for the Ricci tensor under the affine group.

Pseudo-Riemannian manifolds. Geometrically, the case of pseudo-Riemannian constant curvature manifolds is only different in that the separation coordinates are in general not globally defined, but only where the corresponding Killing tensors have real eigenvalues. The algebraic treatment is the same, but yields a much richer variety of separation coordinates, as for other signatures there are many more normal forms for the Ricci tensor under the pseudo-orthogonal group.

Let us condense the above cases in the following programme.

Problem. *Taking the example of spheres as a roadmap, carry through the algebraic geometric approach to a separation of variables for all constant curvature Riemannian and pseudo-Riemannian manifolds. In particular, determine to which extent the corresponding varieties admit tilings by polytopes or carry operad structures which can be used for classification and construction of separation coordinates respectively Stäckel systems. At best, identify these varieties to some known family of moduli spaces arising in a different field of mathematics or theoretical physics.*

For the 3-dimensional hyperbolic space this is an ongoing research project of the author together with Robert Milson.

Non-constant curvature. In comparison to constant sectional curvature manifolds, little is known about separation coordinates on non-constant curvature manifolds and a non-trivial example of a global classification is still lacking. The problem here is already to find good candidates with a non-trivial moduli space. The Killing equation, being an overdetermined partial differential equation, has in general only trivial solutions and even if not, the integrability conditions may only allow trivial solutions.

Motivated by our algebraic approach, we propose three steps for a global classification of orthogonal separation coordinates on a given Riemannian manifold.

Problem. *Find a Riemannian manifold of non-constant curvature with a non-trivial moduli space of separation coordinates and give a* *complete algebraic geometric classification of orthogonal separation coordinates, proceeding as follows.*

Step 1 – Variables: *Identify the isomorphism class of the space of Killing tensors as a representation of the isometry group and construct an explicit isomorphism (as in [MMS04] for constant curvature manifolds). This is the space of variables for the next step.*

Step 2 – Equations: *Translate the Nijenhuis integrability conditions for Killing tensors into purely algebraic equations (as in Chapter 1 for constant curvature manifolds).*

Step 3 – Solution: *Solve these algebraic equations with the help of representation theory and algebraic geometry and interpret the result from an algebraic geometric point of view (as in Chapters 2 and 3 for spheres).*

Apply computer algebra (only) if there is no other way.

At this point we should mention that in dimension two the Nijenhuis integrability conditions are void, so that the variety of Stäckel systems is isomorphic to a $(d-2)$-dimensional projective space, d being the dimension of the space of Killing tensors. In dimension two the above problem is therefore rather uninteresting from an algebraic geometric point of view and reduces to a determination of the space of Killing tensors. Note that Koenigs classified all spaces with $d \geqslant 3$ [Koe72].

4.2 Further notions of variable separation

Complex separation of variables. Separation of variables is often considered on a complex Riemannian manifold right from the beginning. The corresponding theory is essentially the same and our approach also extends to this complex setting without problems. We then deal with the complex sphere

$$\mathbb{CS}^n := \left\{ z \in \mathbb{C}^{n+1} : z_0^2 + \cdots + z_n^2 = 1 \right\} \subset \mathbb{C}^{n+1}$$

and the complex orthogonal group $SO(n+1, \mathbb{C})$ as isometry group. As we did not assume reality throughout their derivation, the algebraic integrability conditions remain the same, just interpreted as algebraic equations in complex variables, i.e. for an algebraic curvature tensor on $\mathbb{C}^{n+1}$. A qualitative difference arises from the fact that complex symmetric forms are not diagonalisable under the complex orthogonal group. As discussed above for hyperbolic space, this will yield different slices for every normal form and a different variety in each of these slices. However, the complex version $\overline{\mathcal{M}}_{0,n+2}(\mathbb{C})$ of the Deligne-Mumford moduli space will only appear for the diagonal normal form, as it is a compactification of the configuration space of $n+2$ ordered distinct points on a complex line modulo the affine group, which correspond to *diagonal* Ricci tensors. We propose the following extension of the above programme to the complex setting.

Problem. *Taking again the example of spheres as a roadmap, extend the algebraic geometric approach to complex separation of variables, by considering the algebraic integrability conditions over the field of complex numbers.*

Notice that we have been deliberately imprecise about the symmetric form g to use in the algebraic integrability conditions. As in the real case, for Euclidean space we need a degenerated one, but non-diagonalisable ones are also imaginable.

R-separation of variables. Instead of looking for a product solution

$$\Psi(x) = \Psi_1(x_1) \cdots \Psi_n(x_n)$$

for, say, the Laplace equation, one can assume the solution to be a product only up to multiplication by an arbitrary function $R(x)$, i.e. to be of the form

$$\Psi(x) = R(x)\Psi_1(x_1) \cdots \Psi_n(x_n).$$

This leads to a weaker notion of separation of variables, called *R-separation of variables*. With some restrictions, the theory is analogous, but with *conformal* Killing tensors in place of ordinary Killing tensors [KM83]. By replacing the isometric embedding $\mathbb{S}^n \subset \mathbb{R}^{n+1}$ with the conformal model $\mathbb{S}^n \subset \mathbb{PR}^{n+1,1}$ of the sphere, we can extend our algebraic approach to the conformal setting, too. In the same way as ordinary Killing tensors on $\mathbb{S}^n$ are described by algebraic curvature tensors in $n + 1$ dimensions, conformal Killing tensors are then described by Weyl tensors in dimension $n + 2$. So we can again translate the Nijenhuis integrability conditions for a conformal Killing tensor to purely algebraic conditions. Nevertheless, a solution of these equations is going to be more difficult even in the simplest non-trivial case $n = 3$, as Weyl tensors form an *irreducible* representation in dimension $n + 2$ greater than 4. On the other hand, being a special case, our solution in the non-conformal case should provide enough information to solve the following problem.

Problem. *Extend the algebraic geometric approach to conformal Killing tensors and R-separation of variables. Start by translating the Nijenhuis integrability conditions for a conformal Killing tensor into purely algebraic conditions on the associated Weyl tensor.*

4.3 Applications

4.3.1 Separable potentials.

Recall that a potential in the Hamilton-Jacobi equation or the Schrödinger equation is compatible with a given system of separation coordinates if and only if it satisfies the Bertrand-Darboux condition (0.9). Such potentials are called *separable*. Note that this condition is linear in V and parametrised by Stäckel systems, i.e. by the moduli space $\overline{\mathcal{M}}_{0,n+2}(\mathbb{R})$ in the case of spheres $\mathbb{S}^n$. If we restrict to square integrable potentials on the sphere, then we can expand V into eigenfunctions of the Laplace operator Δ, which are restrictions of homogeneous functions on the ambient vector space. Moreover, the operator dKd in (0.9) preserves the degree of homogeneity (d lowers the degree by 1, K increases it by 2). Hence we can restrict V to the eigenspaces of Δ, which are finite-dimensional representations of the orthogonal group. In the spirit of our approach, this turns the geometric problem of classifying separable potentials on a sphere into a purely algebraic (even linear) problem on the ambient space.

Problem. *Based on our parametrisation of Stäckel systems on $\mathbb{S}^n$ by the moduli space $\bar{\mathcal{M}}_{0,n+2}(\mathbb{R})$, give a complete algebraic classification of square integrable separable potentials on spheres and, if possible, an algebraic geometric interpretation in terms of sheaves on $\bar{\mathcal{M}}_{0,n+2}(\mathbb{R})$.*

4.3.2 Special function theory

A classification of separation coordinates is nothing but the first step in solving the partial differential equation in question. The next step is to actually *perform* the separation of variables. This will split the partial differential equation in n variables into n ordinary differential equations. Having a smooth parametrisation of separation coordinates at hand should lead to a better understanding of the ordinary differential operators arising in this way, as well as the special functions they define and their interrelation. Before we come to this problem, let us formulate another one, which seems closely related.

Theorem III is both, surprising and unsatisfactory. It is its statement which is surprising, because it relates two seemingly completely unrelated objects: separation coordinates on spheres on one hand and stable genus zero curves with marked points on the other hand. Yet its proof is rather unsatisfactory in that the isomorphism between the corresponding varieties is proven, from an abstract point of view, by comparing the defining algebraic equations and not by directly constructing stable curves with marked points from separation coordinates or vice versa. The latter would be extremely useful for a generalisation of our correspondence.

Problem. *Give a direct construction of a stable curve with marked points from orthogonal separation coordinates (or vice versa), that defines the isomorphism in Theorem III.*

Let us outline a possible answer to this problem. The generic case of separation coordinates on the sphere S^n is the family of elliptic coordinates (0.11). A separation of variables for the Laplace equation in these coordinates yields the *same* equation for each of the n coordinates, namely the (generalised) Lamé equation

$$P\Psi'' + \tfrac{1}{2}P'\Psi' + Q\Psi = 0, \qquad (4.1)$$

where P and Q are polynomials of degree $n + 1$ respectively $n - 1$ [KM86][1]. The zeroes of P are the $n + 1$ parameters of the elliptic coordinates and Q is determined by the separation constants.

A separation of variables in non-generic separation coordinates yields in general a *different* equation in each coordinate. In spherical coordinates on $\mathbb{S}^2$, for example, we obtain a Legendre equation for the latitude and a harmonic oscillator equation for the longitude, resulting in the well known spherical harmonics.

Observe that the Lamé equation is defined on a projective line and has $n + 2$ regular singularities: $n + 1$ at the zeroes of P plus one at infinity. In comparison to this, elliptic coordinates are parametrised in our picture by configurations of $n + 2$ distinct marked points on the projective line, given by the parameters of the elliptic coordinates. That is to say that in the generic case, the curve is the domain of the Lamé equation and the marked points are its singularities. This suggests that the confluent limits of the Lamé equation should not be regarded as a limit where the coefficient P degenerates into a polynomial with multiple roots, but rather a limit where its *domain* – the smooth projective line – degenerates into a singular curve. Then, in a sense to be made precise, the Lamé equation should split into distinct linear differential equations on each irreducible component of the singular curve. For $n = 2$, for example, one should recover the Legendre equation and the harmonic oscillator equation. This is only one motivation for the following problem.

---

[1]Note that the exponent n in Equation (4.5) in [KM86] is a misprint and should be $n + 1$.

Problem. *Give a precise description of all confluent Lamé equations (4.1), based on our parametrisation of separation coordinates on S^n by the moduli space $\bar{\mathcal{M}}_{0,n+2}(\mathbb{R})$. Moreover, exhibit the operad structure on the corresponding ordinary differential operators.*

4.3.3 Generalisations of $\bar{\mathcal{M}}_{0,n+1}(\mathbb{R})$

Let us finally recall that Problem II still remains open. But now we have a lot more information about the variety $\mathcal{S}(\mathbb{S}^n)$ of Stäckel systems on $\mathbb{S}^n \subset \mathbb{R}^{n+1}$. Recall that the Deligne-Mumford moduli space $\bar{\mathcal{M}}_{0,n+2}(\mathbb{R})$ is a compactification of the space $\mathcal{M}_{0,n+2}(\mathbb{R})$ of configurations of $n+1$ distinct ordered points on the real line modulo affine transformations. In our picture these configurations are *diagonal* symmetric forms on $\mathbb{R}^{n+1}$ with distinct eigenvalues and parametrise cofactor systems. Moreover, the action of S_{n+1} on $\bar{\mathcal{M}}_{0,n+2}(\mathbb{R})$ corresponds to conjugation with permutation matrices. The variety $\mathcal{S}(\mathbb{S}^n)$ is now a compactification of the space of symmetric forms with distinct eigenvalues, without imposing diagonality. This yields to the following reformulation of Problem II for spheres.

Problem. *What is the right compactification of the space of symmetric forms with distinct eigenvalues on $\mathbb{R}^{n+1}$ modulo affine transformations? This is a projective variety $\mathcal{S}(\mathbb{S}^n)$ carrying a natural $SO(n+1)$-action, such that if we restrict to diagonal matrices, we recover $\bar{\mathcal{M}}_{0,n+2}(\mathbb{R})$ with the above S_{n+1}-action. Is this variety smooth? Does it have an interpretation in terms of curves with marked points?*

Acknowledgements

My entire research on separation of variables has started with one simple question: "Is the variety of integrable Killing tensors projectively invariant or not?" It was VLADIMIR S. MATVEEV who posed it to me a long time ago, and that is why I would like to thank him first of all. My reasons to thank him are threefold. First, because his insight into the problem from the geometric side and his intuition coming from the theory of integrable dynamical systems have guided me all along the way to find the adequate structures on the algebraic side. Second, because he tolerated the fact that right away from the beginning I ignored his ideas on how to tackle that question. His method definitely would have produced a formal answer within the space of a few weeks. Nonetheless, I hope that the beautiful mathematics I have discovered since then has compensated him for my "lack of obedience". The third reason to thank him is for his patience, especially during the early period, when there were still no results in sight and not even I seemed to be convinced that my endless computations would lead to something useful one day. I am glad that he offered me the scope to take this risk.

The next one who has been crucial for my work is ROBERT MILSON, the first one who fully appreciated my work and recognised its potential. Not only that he approached my unorthodox techniques with an open but critical mind. He even dispelled his doubts about them by writing a computer programme to check everything independently. Probably he now understands my work better than me. The confidence this gave me has been invaluable. I also owe him my gratitude for his spontaneous invitation to present my results at the Dalhousie University in Halifax and his warm hospitality there. After the long period of working in solitude, the inspiring discussions we have maintained on several occasions meant a welcome breath of fresh air to me and gave me an idea of what it means to collaborate, bouncing ideas back and forth. The day after I presented my solution to him, for example, he came up with an even better reformulation of my equations, which also revealed a little mistake in my result. I would finally like to express my thanks for his motivation, which often intervened when I had the impression that everything had already been done. In particular, he encouraged me to attain the link with the classical result.

Speaking of *risk*, I must not omit someone very important from my list: *pure coincidence*. Sometimes it is necessary to be in the right place at the right time and talk to the right person. This is what happened when I met ALEXANDER P. VESELOV in Loughborough. He noticed this tiny link which at the end turned out to be the breakthrough for my approach. We were lucky and everything fitted nicely together. Even better, together with LEONARDO AGUIRRE and GIOVANNI FELDER he had just recently proved exactly the result we needed to prove our conjecture. In my subsequent discussions with SASHA he revealed to me an immense amount of beautiful mathematics. Not to mention the delight to listen to his stories from the history of mathematics. I am still excited about all the intriguing objects that keep emerging from this new perspective on separation of variables, undreamt of when I had set off. I would like to thank SASHA for this contribution that has rounded off my work.

Furthermore I would like to thank STEFAN ROSEMANN, JONATHAN KRESS, CLAUDIA CHANU and GIOVANNI RASTELLI for useful discussions on the subject as well as ANDREAS VOLLMER for proofreading parts of the manuscript.

Last but not least gilt mein Dank in gleicher Weise meiner Familie, ohne die diese Arbeit ebenfalls nie vollendet worden wäre. Bedanken möchte ich mich bei meiner Frau PILAR für ihre moralische Unterstützung sowie die Geduld mit meiner Zerstreutheit durch das fortwährende Nachdenken über mathematische Probleme, bei meinen Eltern WILFRIED & DOROTHEA dafür, dass sie mir seit jeher den Rücken gestärkt und den nötigen Freiraum geschaffen haben und natürlich ganz besonders bei unseren beiden Töchtern ALBA & AITANA für die Abwechslung und die vielen tollen Momente, in denen die Mathematik unwichtig wird.

Bibliography

I find that a great part of the information I have
was acquired by looking something up
and finding something else on the way.

FRANKLIN PIERCE ADAMS *(1881 – 1960)*
New York newspaper columnist

[AFV11]   Leonardo Aguirre, Giovanni Felder, and Alexander P.
           Veselov, *Gaudin subalgebras and stable rational curves*,
           Compositio Mathematica **147** (2011), no. 5, 1463–1478.

[Ben89]   Sergio Benenti, *Stäckel systems and Killing tensors*, Note
           di Matematica **9** (1989), 39–58, Conference on Differential
           Geometry and Topology (Lecce, 1989).

[Ben93]   _____ , *Orthogonal separable dynamical systems*, Differen-
           tial Geometry and its Applications (Opava, 1992), Math-
           ematical Publications, vol. 1, Silesian University Opava,
           Opava, 1993, pp. 163–184.

[BKW85]   Charles P. Boyer, Ernest G. Kalnins, and Pavel Win-
           ternitz, *Separation of variables for the Hamilton-Jacobi
           equation on complex projective spaces*, SIAM Journal on
           Mathematical Analysis **16** (1985), no. 1, 93–109.

[BM03]   Alexey V. Bolsinov and Vladimir S. Matveev, *Geometrical
           interpretation of Benenti systems*, Journal of Geometry
           and Physics **44** (2003), no. 4, 489–506.

[CMS10]   C. Cochran, R. G. McLenaghan, and R. G. Smirnov,
          Equivalence problem for the orthogonal webs on the sphere,
          arxiv:1009.4244v1 [math-ph], 2010.

[CMS11]   _____, *Equivalence problem for the orthogonal webs on
          the sphere*, J. Math. Phys. **52** (2011), 1–22.

[Cra03]   Michael Crampin, *Conformal Killing tensors with vanish-
          ing torsion and the separation of variables in the Hamilton-
          Jacobi equation*, Differential Geometry and its Applica-
          tions **18** (2003), no. 1, 87–102.

[Dev99]   Satyan L. Devadoss, *Tesselations of moduli spaces and the
          mosaic operad*, Homotopy invariant algebraic structures
          (Baltimore, MD, 1998), Contemporary Mathematics, vol.
          239, American Mathematical Society, Providence, RI,
          1999, pp. 91–114.

[Dev09]   _____, *A realization of graph associahedra*, Discrete
          Mathematics **309** (2009), no. 1, 271–276.

[DHMS04] R. J. Deeley, J. T. Horwood, R. G. McLenaghan, and R. G.
          Smirnov, *Theory of algebraic invariants of vector spaces
          of Killing tensors: methods for computing the fundamental
          invariants*, Symmetry in nonlinear mathematical physics,
          Pr. Inst. Mat. Nats. Akad. Nauk Ukr. Mat. Zastos., 50,
          Part 1, vol. 2, Natsīonal. Akad. Nauk Ukraïni Īnst. Mat.,
          Kiev, 2004, pp. 1079–1086.

[DJS98]   Michael W. Davis, Tadeusz Januszkiewicz, and Richard A.
          Scott, *Nonpositive curvature of blow-ups*, Selecta Mathe-
          matica New Series **4** (1998), no. 4, 491–547.

[DR01]   Satyan L. Devadoss and Ronald C. Read, *Cellular struc-
          tures determined by polygons and trees*, Annals of Combi-
          natorics **5** (2001), no. 1, 71–98.

[EHKR10] Pavel Etingof, André Henriques, Joel Kamnitzer, and Eric M. Rains, *The cohomology ring of the real locus of the moduli space of stable curves of genus 0 with marked points*, Annals of Mathematics. Second Series **171** (2010), no. 2, 731–777.

[Eis34] Luther P. Eisenhart, *Separable systems of Stäckel*, Annals of Mathematics. Second Series **35** (1934), no. 2, 284–305.

[FKWC92] S. A. Fulling, R. C. King, B. G. Wybourne, and C. J. Cummins, *Normal forms for tensor polynomials: I. The Riemann tensor*, Classical Quantum Gravity **9** (1992), 1151–1197.

[HMS05] J. T. Horwood, R. G. McLenaghan, and R. G. Smirnov, *Invariant classification of orthogonally separable Hamiltonian systems in Euclidean space*, Comm. Math. Phys. **259** (2005), 670–709.

[Kal86] Ernest G. Kalnins, *Separation of variables for Riemannian spaces of constant curvature*, Pitman Monographs and Surveys in Pure and Applied Mathematics, vol. 28, Longman Scientific & Technical, Harlow, England, 1986.

[Kap93] Mikhail M. Kapranov, *The permutoassociahedron, Mac Lane's coherence theorem and asymptotic zones for the KZ equation*, Journal of Pure and Applied Algebra **85** (1993), no. 2, 119–142.

[KM83] Ernest G. Kalnins and Willard Miller Jr., *Conformal Killing tensors and variable separation for Hamilton-Jacobi equations*, SIAM Journal on Mathematical Analysis **14** (1983), no. 1, 126–137.

[KM86] _____, *Separation of variables on n-dimensional Riemannian manifolds. I. The n-sphere S_n and Euclidean n-space R_n*, Journal of Mathematical Physics **27** (1986), no. 7, 1721–1736.

[KMR84]   Ernest G. Kalnins, Willard Miller Jr., and G. J. Reid,
          Separation of variables for Riemannian spaces of constant
          curvature. I. Orthogonal separable coordinates for S_{nC}
          and E_{nC}, Proceedings of the Royal Society London **394**
          (1984), no. 1806, 183–206.

[Knu83]   Finn F. Knudsen, *The projectivity of the moduli space*
          of stable curves. II, III, Mathematica Scandinavica **52**
          (1983), 161–212.

[Koe72]   Gabriel X. P. Koenigs, *Sur les géodésiques a integrales*
          quadratiques, Le cons sur la théorie générale des surfaces
          (Jean G. Darboux, ed.), vol. 4, Chelsea Publishing, 1972,
          pp. 368–404.

[Lam37]   Gabriel Lamé, *Sur les surfaces isothermes dans les*
          corps homogènes en équilibre de température, Journal
          de Mathématiques Pures et Appliquées 1re Série **2** (1837),
          147–188.

[Lee89]   Carl W. Lee, *The associahedron and triangulations of*
          the n-gon, European Journal of Combinatorics **10** (1989),
          no. 6, 551–560.

[Lod04]   Jean-Louis Loday, *Realization of the Stasheff polytope*,
          Archiv der Mathematik (2004), no. 83, 267–278.

[MHPS12] Folkert Müller-Hoissen, Jean Marçel Pallo, and James D.
          Stasheff (eds.), *Associahedra, Tamari lattices and related*
          structures, Progress in Mathematics, no. 299, Birkhäuser,
          2012.

[MM10]   Vladimir S. Matveev and Pierre Mounoud, *Gallot-Tanno*
          theorem for closed incomplete pseudo-riemannian mani-
          folds and applications, Annals of Global Analysis and
          Geometry **38** (2010), no. 3, 259–271.

[MMS04] R. G. McLenaghan, Robert Milson, and R. G. Smirnov, *Killing tensors as irreducible representations of the general linear group*, Comptes Rendus de l'Académie de Sciences Paris **339** (2004), 621–624.

[Neu59] Carl Neumann, *De problemate quodam mechanico, quod ad primam integralium ultraellipticorum classem revocatur*, Journal für die reine und angewandte Mathematik **56** (1859), 46–63.

[Nij51] Albert Nijenhuis, X_{n-1}-*forming sets of eigenvectors*, Proceedings of the Koninklijke Nederlandse Akademie van Wetenschappen **54** (1951), 200–212.

[OEISa] The Online Encyclopedia Integer Sequences, *Sequence A001190*, published electronically at http://oeis.org/A001190.

[OEISb] ———, *Sequence A032132*, published electronically at http://oeis.org/A032132.

[Ole50] M. N. Olevskiĭ, *Triorthogonal systems in spaces of constant curvature in which the equation $\delta_2 u + \lambda u = 0$ admits separation of variables*, Mat. Sbornik N.S. **27(69)** (1950), 379–426.

[Sch10] Konrad Schöbel, *Algebraic integrability conditions for Killing tensors on constant sectional curvature manifolds*, arxiv:1004.2872v1 [math.DG], 2010.

[Sch12] ———, *Algebraic integrability conditions for Killing tensors on constant sectional curvature manifolds*, Journal of Geometry and Physics **62** (2012), no. 5, 1013–1037.

[Sch14] ———, *The variety of integrable Killing tensors on the 3-sphere*, SIGMA **10** (2014), no. 080, 48 pages.

[Sch15] Konrad Schöbel, *Nijenhuis integrability for Killing tensors*, 2015, arxiv:1502.07516 [math.DG].

[ST69]   Isadore M. Singer and John A. Thorpe, *The curvature of 4-dimensional Einstein spaces*, Global Analysis (Papers in Honor of K. Kodaira), University of Tokyo Press, Tokyo, 1969, pp. 355–365.

[Stä91]   Paul Stäckel, *Über die Integration der Hamilton-Jacobischen Differentialgleichung mittelst Separation der Variabeln*, Habilitationsschrift, Universität Halle, Halle, 1891.

[Sta63]   James D. Stasheff, *Homotopy associativity of H-spaces. I, II*, Transactions of the American Mathematical Society **108** (1963), no. 2, 275–312.

[SV15]   Konrad Schöbel and Alexander P. Veselov, *Separation coordinates, moduli spaces and Stasheff polytopes*, Communications in Mathematical Physics **337** (2015), no. 3, 1255–1274.

[Vil65]   Naum Ya. Vilenkin, *Special functions and group representation theory*, Nauka Publ. Comp., Moscow, 1965, (in Russian).

[VK93]   Naum Ya. Vilenkin and A. U. Klimyk, *Representation of Lie groups and special functions. Volume 2: Class I representations, special functions, and integral transforms*, Mathematics and its Applications (Soviet Series), vol. 74, Kluwer Academic Publishers Group, Dordrecht, 1993.

[VO04]   Anatoliĭ M. Vershik and Andreĭ Yu. Okun′kov, *A new approach to representation theory of symmetric groups. II*, Rossiĭskaya Akademiya Nauk. Sankt-Peterburgskoe Otdelenie. Matematicheskiĭ Institut im. V. A. Steklova. Zapiski Nauchnykh Seminarov (POMI) **307** (2004), 57–98, English translation in: Journal of Mathematical Sciences (New York) **131** (2005), 5471–5494.

Printed in the United States
By Bookmasters